The Illustrated History of
TRACT**O**RS

From pioneering steam power to today's engineering marvels

Robert Moorhouse

CHANCELLOR
PRESS

A QUANTUM BOOK

This edition published in 2000 by Chancellor Press
An imprint of Bounty Books, a division of
Octopus Publishing Group Ltd
2-4 Heron Quays
London
E14 4JP

ISBN 0-75370-368-8

QUMTOR

This book was produced by
Quantum Publishing
6 Blundell Street
London N7 9BH

Printed and bound in Singapore by
Star Standard Industries Pte Ltd

CONTENTS

Introduction

When I was two months old, the bulldozers arrived on my parent's farm to clear most of it for mining. Tractors have featured in my life ever since. Actually my first encounter with a tractor was disastrous. When dad's first Fordson N was started up to drive it off the delivery truck, I screamed so loud, I almost drowned the engine noise!

When growing up, most boys played football and cricket; but not me. My friends and I watched the bulldozers and scrapers and compared tractors on the village farms – Fordson, David Brown and Nuffield. Trips to the local market revealed other makes such as International, Massey Harris and Ferguson.

In the early 1950s my friend's grandfather died and the farmhouse was cleared out. Amongst the rubbish were piles of *Farmers Weekly* and *Stock Breeder* magazines, so one wet day during school vacation, and having nothing else to do, my friend and I started looking at them. A whole new world of tractors opened up to us. Wartime issues showed John Deere, Minneapolis Molines, and other American makes, so we started saving the pictures. Better weather and other interests came along but I was firmly hooked on tractors and I still have some of the original cuttings. I wish I had kept the magazines as well.

When I was at school I won a class prize which was a book of my choice. I asked for *Tractors on the Farm*, by H J Hine but the teacher was so baffled by my choice that he referred me to the principal teacher who fortunately took the view that my unusual choice meant that I would appreciate it. That well-thumbed book is still on my bookshelf today.

Working on my parent's farm and then on my own farm provided plenty of magazines and dealers' brochures. Correspondence with the late F Hal Higgins and Elmer J Baker amongst others opened up the American scene to me. When tractor preservation came along I joined the infant Vintage Tractor club. Meeting the late Charles Cawood, who was probably one of the first tractor historians, provided the impetus to really study tractor history and we spent endless hours discussing the subject.

It is said that everyone has a book in them and a telephone call in late 1995 asking if I would write one, provided the opportunity to put that statement to the test.

There are many tractor books available to the enthusiast so why write another one? This book is how I see the global history of the tractor, it may not be everyone's view in which case I hope the reader will be tolerant. In any industry selling to world markets, the companies involved copy, modify and adapt their products to local conditions, so it is with tractors. The American farmer's ideal tractor may not suit his Swedish counterpart and the British requirement is different from the Australian. They all come from a common origin, the Charter of 1889, and I hope this book goes some way to showing how they all evolved into today's wonderful machines.

Robert Moorhouse

The Origins of a Species

From the early days of agriculture to the harnessing of steam power and the development of the internal combustion engine. As with any invention, machines to help on the farm were being developed by different people in different parts of the world: what emerged were a number of forerunners of the tractor as we know it today.

From animals to machine. Modern-day
farming now relies on the tractor for a
multitude of tasks and is certainly a long
way from the machine's humble origins.
Here, corn is being harvested for silage.

The Origins of a Species

There are few places on the globe where tractors cannot be seen in one form or another. They work on farms worldwide and, in their industrial guise, work the world's mines and forests. They are also used in the construction industry and factories which makes them a truly universal machine.

Tractors have been to the South Pole and are in use on remote islands; Tristan da Cunha has tractors but no other vehicles. They come in all shapes and sizes, from tiny models for use in gardens to giants of 1000hp. Most are wheeled but crawlers still have a place for many applications.

How they all came about is the subject of this book. The word tractor is generally considered to be of recent origin, however, like a lot of words it has Latin roots. *Tractorius* was used for the act of drawing or pulling in a mechanical context. By medieval times, when Latin was still in regular use, the word *tractor* appears, still in the same context. With the rise of steam traction engines in the 19th century, the word returned to use and was often applied to these engines.

The earliest modern usage according to R B Gray in his book *Development of the Agricultural Tractor in the United States*, was in a patent issued in 1890 by the United States Patent Office to a Mr G H Edwards of Chicago; however there is no record of this machine ever being built. The *Oxford English Dictionary* gives the following origins:

1 The 1903 *Motor Manual* published in Great Britain notes that "Rhodesia has

Tractors have come a long way from their humble origins. This modern-day Case tractor and forage harvester is making silage.

appealed to motor manufacturers to supply motor wagons or tractors for use in hilly country".

2 A supplement to the *Scientific American* for November 4th 1905 states "At a recent show of the British Royal Agricultural Society great interest was centered on the Scott motor tractor ... The motor in this tractor is a 24hp ... standard Astor Engine".

The Petter Company of Yeovil, England had, in 1903, exhibited a tractor clearly labelled PETTER'S PATENT AGRICULTURAL TRACTOR.

There is a popular story in the United States that W H Williams of the Hart Parr Company of Charles City, Iowa, invented the name while writing an advertisement in 1906. The story goes

that while pondering what to put on an advertisement, the words 'Gasoline Traction Engine' seemed too clumsy and he was inspired to write 'tractor' to replace them. Whether this really took place we do not know, but as the old Italian proverb says "*Se non e vero, e ben trovato*" which translates as "it may not be true, but it is a good story." Today, with minor variations in spelling, the word tractor is almost as universal as the machine it describes.

So much for the name, from where did the machine originate?

From the dawn of the human race, people were hunters and gatherers until someone learned how to domesticate animals and grow plants for food.

Agriculture had arrived and with it the ability to settle in one place. Eventually settlements arose that enabled the pursuit of things other than food collecting. Farmers were thus charged forever with the responsibility of feeding an emerging urban population. For centuries they did this with the aid of animal power to supplement their own. Cattle, including water buffalo, camels and later horses, donkeys and mules, bore the brunt of this in various parts of the world, mostly for plowing as few other implements existed.

By the time of the Renaissance there was a pressing need to improve agriculture. Jethro Tull invented the seed drill which, by setting the seed in rows, enabled hoeing to take place: weeds having always been the farmer's enemy. Threshing the grain also came in for improvement. After a false start trying to make mechanical flails, notably by Christopher Pohlem in the 1720s, a Scotsman, Andrew Meikle, laid down the basic method of threshing grain using a fast revolving cylinder fitted with bars that rubbed the grain through a grating held close to the cylinder. The grain fell on

Ancient farm power was largely provided by animals to aid that of the farmer. Cattle - one of the earliest domesticated animals - was frequently used to plow the fields. Later, horses and mules helped man in his endeavors.

to a reciprocating shaker which separated out most of the grain from the chaff. Wind, water and muscle, both animal and human, still formed the power source for all these machines. What was needed was a mechanical power source.

Steam power

Numerous people had been experimenting, with some success, with steam power, mainly for pumping water from mines or to power the fountains in the gardens of the country houses that were rapidly becoming fashionable in Britain and the rest of Europe. Thomas Newcomen, a mining equipment supplier, took an interest in the steam pump concept and in 1712 built a successful steam pump engine. From this basis the steam engine was gradually improved, to a large extent by James Watt, until in 1769 Nicholas Cugnot, a French farmer's son who had joined the army, built a steam artillery tractor. This was the first machine designed to tow anything: the tractor was in conception. However, as the army declined to adopt the idea, so the steam tractor languished for almost a century.

With the invention of steam came an early glimmer of the embryonic tractor. However, in farm work, most steam engines were stationary, being used to aid laborious jobs such as threshing.

Steam power rapidly made headway with the invention of the railway locomotive, and eyes were once again cast to the farming scene. In Britain and Europe with their small farms, the main use was for hauling and powering threshing machines, although limited success was had in applying it to field work. One notable method was the cable plowing system using two engines, one on each side of the field, with the plow or other implement pulled to and fro by wire ropes on winches beneath each engine. John Fowler of Leeds, England, was the main advocate of this method which lasted until the early 1930s. The last ones were diesel driven sets built by McLarens, also of Leeds, for use in the Sudanese cotton fields.

In North America, conditions were very different to those in Europe. Here the need was to plow up the prairies for wheat to feed the growing population as well as for export. Once the traction engine had been brought to a workable state, numerous companies turned out steam traction engines by the thousand. Engines capable of pulling up to 20 furrow plows by direct traction were built for breaking the prairie sod. Harvesting was still a problem, the binder being essentially a small machine, and multiple hitches were cumbersome in use. In drier parts of the States, the combine harvester appeared on the harvest field and by its size lent itself to be steam hauled. However, cultivation and seeding remained the province of the horse in most areas. Steam engines were only employed on a very large scale farm.

Farmers in South America, particularly in Argentina, South Africa, Australia and the growing British Empire, faced similar problems to their American cousins. Not having had time to evolve their own industries, they relied on imports from Britain, Europe and the United States of America. Worldwide farming was experiencing a shake up but it was not without its problems; workers and farmers could be very conservative and there was resistance from some people. Just as the Luddites, led by the mythical General Ludd, had smashed up factory machinery, so gangs of farmworkers in Britain went to farms to burn stacks and any threshing machines they could find. They saw this as a threat to their livelihood.

Farmers too, especially in Britain and Europe, were very resistant to the attempts to work their land by steam on the grounds that it was compacting the soil. This was at least true in the wetter, heavier soils of Western Europe. Several steam companies made attempts, right up to the 1920s, to produce a small steam tractor for direct field work but sadly, they all failed.

The history of the traction engine is well written about, and numerous books exist for anyone who wishes to know more than this brief mention.

Internal combustion engine

While this was taking place, men in workshops all over Europe were investigating ideas for an alternative to the steam engine with its huge appetite for fuel and water. Men had long dreamed of an "engine" and in the 1680s two people actually made one using gunpowder as fuel. Christiaan Huygens, a Dutchman, and Denis Papin, a Frenchman, experimented with these "engines" for a while but they were too erratic to use. They could be considered to be the founding fathers of the internal

An example of early cable plowing, where a plow was pulled by wire ropes on winches to and fro across the field. This method - which lasted in various forms until the early 1930s - appears rather clumsy by today's standards.

Portable fuel

In the United States of America a certain Mr Drake drilled an oil well in 1859 and started the oil industry which in turn provided the 'portable fuel' needed for a vehicle powered by an internal combustion engine. The initial use of the oil was for fuelling lamps and the lighter fractions were burnt off as waste. As the industry progressed, these became available as liquids and the fuel of the future came into being.

combustion engine. Fuel was a problem and it was not for another 100 years, with the discovery of coal gas, that anyone was more successful. This time the experimenters were English. In 1790, John Barber built an engine fuelled by coal gas and four years later Robert Street used turpentine for fuel, which makes his discovery possibly the earliest liquid fuelled internal combustion engine.

Progress in engineering in general was helped by the big improvements made to machine tools. Maudslay's lathe in Britain and Eli Whitney's milling machine in America are two examples that demonstrated the precision engineering needed by the internal combustion engine to move it towards being a workable alternative to steam. The piston needed to be a tight fit but easy to move in the cylinder. By the mid 19th century things were progressing rapidly. Etienne Lenoir patented a gas engine in France and put it into production. Licensed production took place in Britain by the Reading Iron

Works. Another Frenchman, Alphonse Beau de Roches, formulated the four-stroke principle in 1862 but did not put it into use.

The next and most significant entrant on the scene was the German, Nikolaus August Otto. Otto, intrigued by Lenoir's gas engine success, built two under licence and then set about improving on the design to make them more efficient. He teamed up with a fellow engine experimenter Eugen Langen and formed a company to build the new design. These engines worked reasonably well but were extremely noisy in operation. After winning a gold medal at the 1867 World Exhibition in Paris, they set about a further redesign and in 1876 came out with a four-stroke gas engine so much quieter than its predecessor, it was nicknamed the 'Otto Silent'. This employed the basic engine cycle of operation used today in that it introduced:

1 induction of the air and fuel mixture
2 compression of the mixture

3 power from the burning of the fuel
4 exhaust of the burnt gases and so back to induction.

By 1880, the company founded by Otto and Langen became known as the Gasmotoren Fabrik Deutz AG and among its staff were Gotlieb Daimler and Wilhelm Mayback. These two engineers foresaw the possible adaptation of the four-stroke cycle to liquid fuel and hence its ability to power a motor vehicle. In 1882 they set up independently from Deutz to pursue their ideas. The following year Daimler and Mayback built a four-stroke engine using a petroleum-based fuel called Benzin. By late 1885 this engine was running at the unheard speed of 900rpm and powered their first vehicle, a motor bicycle.

Quite independently another German, Karl Benz, a manufacturer of gas engines, had built a four-stroke engine very similar to the Daimler and Mayback and used it to power a tricycle. The internal combustion engine now had wheels and was ready to go places.

In a sense, the internal engine did it initially without wheels, for in 1886 it crossed the Atlantic Ocean to the United States of America. This country more than any other was to exploit the internal combustion engine to its limits. (Rather confusingly, this type of engine is called a gas engine, short for gasoline, but true gas engines, those that operated with gaseous vapour, are very rare.) The Otto engines were brought in by Schleicher, Schumm and Company of Philadelphia who then built them under licence.

By the late 1880s, practically all the components needed to build a workable tractor had or were about to come into existence. Rudolph Ackerman, a German resident in Britain, had formulated his

geometrically correct steering system and a pioneer cyclist called James Starley built a modern style differential in 1877. Numerous other people invented or improved items such as the gearbox, friction clutch and very importantly better carburettor and ignition systems. The day of the tractor was approaching, it just needed the right conditions.

In Britain, farming was a depressed industry: wheat was imported from North America and from 1880 the importing of frozen meat from Australia left farmers in no position to embrace the possibilities of the gasoline engine in any form. But in the United States of America, things were better. The aftermath of the Civil War had passed and the west was being opened up. In 1869, the railroad crossed the continent. The rapid expansion of the railroad system enabled farmers to send their produce to market and bring in the

things they needed. Steel plows, seed drills and implements of all kinds were in big demand as farmers sought to feed the growing population.

As the century progressed, the weaknesses of the steam engine became apparent. Fuel and water in huge quantities were needed and the further west, the scarcer these became. The weight also caused problems. For initial prairie plowing, the sod would support the engine, but after a few years when this had rotted away it was necessary to use horses or mules. But they took up land and - like the steamer - a lot of labor, itself a scarce item.

The conditions were now right for the next step and all the required items were available. The foundations of the tractor industry were now laid and it just needed someone to build on them. These builders were not long in coming.

The 'Farmer's Engine' - a steam traction engine in the fields of Cambridgeshire, England around 1870.

The Formative Years

By the turn of the century, the pioneers of tractor design and development were producing their machines in numbers for sale. Names such as Otto, Case, Saunderson, Hornsby, Deutz, the International Harvester Company, Hart and Parr all entered the tractor arena.

A 1903/4 model of the Ivel tractor. This small, easy to use machine, looked to the fledgling automotive industry for its design rather than the earlier steam engines.

The Formative Years

I n 1889 in South Dakota, an area still thought of as the Wild West, history was made on the wheat ranch of Mr L F Burger when the first tractor went to work driving a threshing machine. Sources differ as to whether it was built to his order or whether he bought it from the builder. In either event, it does not really matter, Mr Burger had the courage of his convictions that the steam engine could be replaced by the internal combustion engine.

The builder of the first tractor was John Charter, who had been successfully building gasoline engines in Stirling, Illinois for several years, patenting numerous improvements as he did so. By the simple expedient of using one of his large 25hp single cylinder stationary engines on the running gear of a Rumely steam traction engine he made a workable tractor. The tractor was sufficiently successful so that about six more were built and sold to farmers in South Dakota. Strangely, Charter made no more, however the tractor's success must have been noted by others because in 1892 three other tractor makers appeared, one of which, Case, would last to the present day.

Case unfortunately used a very odd engine built to the design of William Paterson. This used a single cylinder with two pistons and a complicated ignition system which, together with a surface evaporation carburettor, proved to be a failure when used in a tractor. Case decided that the internal combustion engine tractor belonged to the future and

went back to steam; Case's company was the largest steam traction engine builders in the USA at this time.

Another early maker was John Froelich. One of the first tractor sales catalogs was issued by him and it makes interesting reading. Froelich claimed to have thought up the idea in 1886 while he worked as a threshing contractor in the wheat lands of the Northwest. In 1892 he bought a Van Duzen 16hp single cylinder vertical engine, built his own transmission and steering system and mounted them on a Robinson steam engine running gear. Such was his confidence that he then bought the

largest Case threshing machine plus cooking and sleeping wagons and sent the outfit by rail to Langford, South Dakota where hundreds of farmers turned out to witness its arrival. Working in temperatures between 100°F and -3°F for a 50 day run, over 62,000 bushels of wheat were threshed without any breakdowns. As a result of all this a group of business men in Waterloo, Iowa, together with Froelich, formed the Waterloo Gasoline Traction Engine Co. "with an abundance of capital".

Froelich's catalog - probably issued in late 1893 - gives numerous testimonials to the tractor's performance and its

One of the earliest tractors manufactured by Case in 1893.

advantages over steam. Many farmers, however, were still unconvinced and in 1895 the word Traction was dropped from the company name. Waterloo continued to make stationary engines until re-entering the tractor field in about 1912.

In the run up to the 20th century, several other makes of tractor appeared in the USA with varying success. Two former employees of Schleicher, Schumm and Company, who founded their own company, The Dissinger Brothers, built a prototype called 'The Capital' and experimented with it for a number of years before making any more. The following year the Sterling or Hockett tractor was offered for sale without any success. This was said to be the first tractor advertised for general sale. But production really started in 1898 when the Huber Company of Marion, Ohio, bought the Van Duzen Engine Company and proceeded to make 30 tractors -based on the Van Duzen prototype - the same year. There is a possibility that one of these later became the first tractor to pull a plow. However, there is also a sketch showing a Froelich tractor plowing, so the matter is at present unresolved.

By 1894, Scheicher, Schumm and Company had become the Otto Gas Engine Co when they built their first tractor. Fourteen more were built before 1900. After a promising start they built small numbers until about 1914 when production ceased.

Kinnard-Haines from Minneapolis started building tractors in 1897 and by 1899, now called the Flour City, they had built about 30. In his book *Farm Tractors in New Zealand* Richard Robinson notes that four Kinnard-Haines tractors were allegedly sold there in 1902. If this is correct then Kinnard-Haines was the first American tractor company to go into the export business.

In 1899, S S Morton built a tractor with friction drive. Compared with other tractors of the period it was a very compact design. He then went on to sell tractor chassis and friction drive units on which buyers could install the engine of their choice.

Although not strictly tractors, both Deering and McCormick built self-propelled mowers in the late 1890s. They made no impact in the USA but many years later this style of light tractor mower would be extremely popular in the mountainous parts of Europe.

All this time, the American tractor builders looked to replace the steam engine. In Europe, ideas took a different path, they looked to replace the horse.

Europe

Around 1896 in Germany, home of the internal combustion engine, a tractor was built by Adolf Altman, an engine manufacturer. This used a single cylinder, horizontal hopper-cooled engine with chain drive to the wheels. There was no differential, so for turning, one wheel was disconnected. This design was not successful and Altman sold his company soon after. One interesting point emerges from this: he referred to his machine as a *trakteur*, possibly the first time this word was used outside the English language.

It was the work of another German whose engine would ultimately prove to be the best tractor engine of all. His name was Rudolph Diesel. Diesel's engine used very high compression to ignite the fuel and so eliminate the troublesome early electrical systems. Akroyd-Stuart used a low compression and a hot bulb, initially heated by blow lamp, to achieve the same ideal of no electrical ignition system.

Two Englishmen were also at work on tractor designs although neither were built until after the turn of the century. They were Dan Albone and Herbert Saunderson, both from Bedfordshire. It also seems that Petter's of Yeovil,

Only fifteen Otto tractors were built before 1900. This one dates from 1896.

England were also at work on a tractor A British company that did build and sell tractors in the 1890s was Richard Hornsby and Sons of Grantham, Lincolnshire, who built Akroyd-Stuart kerosene engines under licence. In 1896 they built their first Patent Safety Oil Traction Engine using one of these engines, and made the first sale of a tractor in Britain. The following year half a dozen were exported to Australia. Tractors were starting to spread around the world.

Hornsby-Ackroyd engine built just prior to the turn of the century and shown here beautifully restored. It is driving a threshing machine.

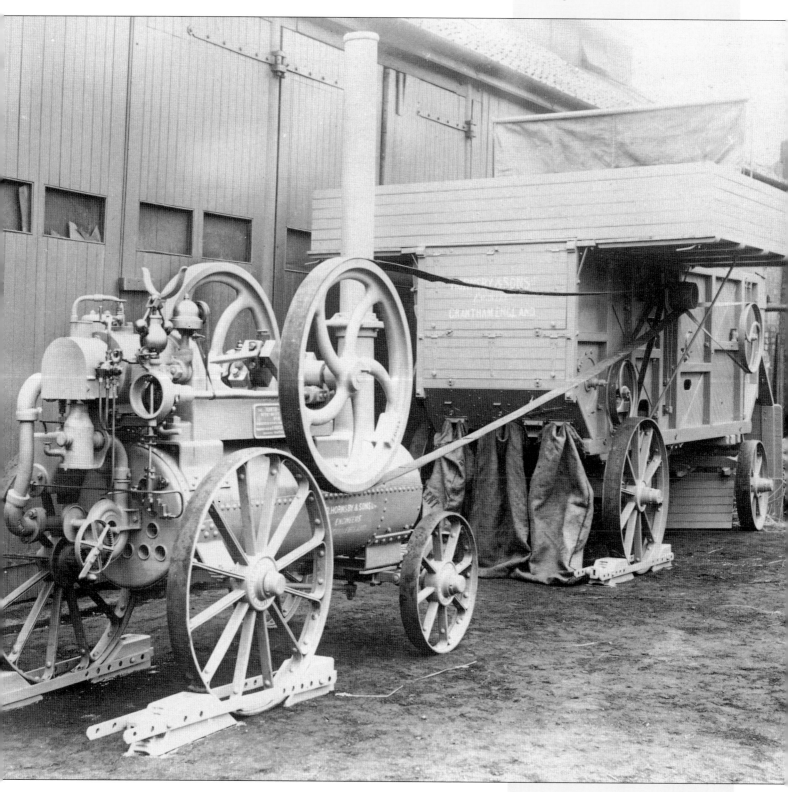

Above: The original Ivel agricultural motor tractor now housed in the Science Museum in London, England.
Right: A 1903 Ivel, still in working order fifty years later.

Into the 20th century

The first years of the new century produced some pointers to the future. Dan Albone, bicycle manufacturer and engineer, built his first Ivel tractor, named after the local river. Dan, having no steam background, and not seeking to replace the traction engine looked more to the fledgling automotive industry for his design. The result was a small, easy to use machine that would do just about any belt or field work, except rowcrop work, on the average British farm. Production started and by 1906, when Dan's untimely death stopped the development of the Ivel, tractors were being exported to many countries. In response to a request an orchard model was made and sent to Tasmania, an early example of a tractor

for a special application. But by 1916 the company had lost its impetus and production ceased. They then took an agency to sell the Hart Parr Little Devil in Britain. A 1917 publication states that it was being marketed by Britain's oldest tractor makers. The contrast between the two tractors is stark; from an advanced design to an odd single wheel drive model with no reverse gear, except by stalling the engine so it would run backwards. It is doubtful if many were sold in Britain.

In rapid succession, several tractors of advanced design appeared in Britain. The Scott first appeared in 1903 and featured a power-driven rear cultivator with seed drill on top, the first integrated tractor and powered implement. Ransome's of Ipswich produced their Agricultural Petrol

Ransome's Agricultural Petrol Motor Tractor sadly did not last long. This one, built in 1904, cost around 450 English pounds to buy.

Military use

Apart from farmers and threshermen, there was another group of people who needed to tow loads off the roads, the Army. It had not escaped the notice of the British Army that the internal combustion engine had arrived and in 1903 they organised a trial for a suitable tractor. The £1000 first prize went to Hornsby's of Grantham. Military tractors are really outside the scope of this book, but one tractor needs to be described. David Roberts, Managing Director of Hornsby's, had patented a crawler track and in 1906, at the request of the Army, Hornsby's fitted a set to one of the 1903 tractors. This monster crawler weighing 23 tons and producing over 70hp, was the world's first full crawler driven and steered by its tracks using compressed air operated brakes. The soldiers who saw the trials called it the Caterpillar and the name stuck. By 1909 Hornsby's had produced four very modern looking full crawlers called Little Caterpillars for the Army. There was no further interest and the patent was sold to Holt in America in 1912.

Right: The Hornsby military tractor of 1905 after conversion from wheels to tracks.

Above: The Hornsby Little Caterpillar, manufactured in 1909, usefully engaged by the British Army just prior to the outbreak of World War I.

Left: The original Hornsby single-cylinder prototype crawler of 1904.

Sharp's original tractor with its mid-mounted power-driven mower built early this century.

their Agricultural Gasoline Motor Tractor with a 4 cylinder in-line engine and three speed and reverse gearbox. At £450 it, like the Scott, did not sell and soon left the market. A similar model came from Sharp complete with mid-mounted power-driven mower. Several other companies produced prototypes or small numbers during this period. Other notable features of these tractors were the use of Ackerman steering and high tension magnetos - the high tension magneto having been invented earlier in 1895 by Frederick Simms of London.

Petter's, who had been experimenting with tractors, produced in 1903 their Petter's Patent Agricultural Tractor with forward control, a hot tube ignited kerosene engine and three forward and reverse gearbox. The design may indicate that Petter's expected it to be fitted with a rear platform for haulage purposes. If so, they were about 65 years too soon in commercial terms.

Marshall's and Daimler's both looked to the British Empire and Colonies for their sales. They produced big prairie style tractors similar to the Americans. Marshall's appeared at the 1908 Winnipeg Trials and sold quite well in Canada and Australia. The Daimlers went up to 110hp, the largest in the world at that time.

Herbert Saunderson took the horse replacement idea literally and made

Above: The Hornsby/Foster steam caterpillar of 1912 supplied to the Northern Light, Coal & Power Company in the Yukon.
Left: The York 12 and 16hp double cylinder traction engine of 1906 – an early American attempt at a small tractor.

several two wheeled tractors that could be attached to horse carts and implements. He also made several all-wheel drive tractors of various configurations, many of which were exported. He then settled on a more conventional style four wheeler, albeit with front seat and rear cooling water tank, before moving them round to front radiator and rear driving position.

On the European Continent, the first signs of motorized farming appeared. In fact, in 1896, a machine that resembled a cross between a rotary cultivator and a plow had been built in Germany as well as Altman's tractor. Another one came out in 1907, forerunner of the Stock motor plows who later made a variety of models. The motor plow design became very popular and initially put ordinary tractors very much into second place.

Deutz, who today are major tractor manufacturers, brought out a tractor and a motor plow in 1907. The basic tractor unit of the motor plow had four wheel drive and a four furrow plow at each end with winch lift, a very advanced machine at the time. Some 70 years later this design would be revived by tractor companies and tried again with rubber tires, a diesel engine and hydraulic lifts.

A number of other companies were also at work on a farm tractor suitable for European conditions, most of them motor plow designs for the plains of the German Empire and its close ally the Austro-Hungarian Empire. By the outbreak of the Great War, models of up to 80hp with six furrows were available from companies like Stock and Deutsche Kraftpflug, both of Berlin, Hanomag from Hanover and Pohl. Pohl from Grossnitz also built large 60hp conventional style tractors with mounted cable lift plows of up to six furrow capacity, quite a large machine. The powered rotary cultivator also received quite a lot of experimental work form Lanz and others. The Lanz with an

Produced from 1912, this French Lefebvre tractor housed a four-cylinder engine which developed 35hp. It had a multi-disk clutch, a two-speed gearbox and was able to raise its tracks and run like a wheeled tractor.

80hp, four cylinder engine and three wheel design was later improved and available with four wheel conventional layout.

The French produced several prototypes from the De Souza of 1904 to the important Gougis tractor. One of the most useful items on a modern tractor is the independent power take off shaft. Gougis invented it in 1907 when he built a tractor to power a converted McCormick binder. A three wheeled tractor, it had a four cylinder in-line engine with a p.t.o. shaft at the rear of the tractor directly driven from the engine through a separate clutch and hence by a universally jointed shaft through to the binder. Later shown at several exhibitions, it was seen by a representative of the International Harvester Company. Gougis did not put his machine into production in spite of its promising design.

Row-crop work, later to become an important tractor function, was first carried out by another Frenchman called Bajac. His tractor, again only in prototype form, was powered by a two cylinder engine and used a rear mounted cultivator with hand lift.

In the hope of encouraging French tractor makers, a trial was held in 1913 at the National Agricultural College at Grignon near Paris. Thirteen machines were present for a series of trials both compulsory and voluntary. The variety of tractors was wide, three had attached plows, another three had powered rotary cultivators and two used winches to pull the implements. The largest machine present was the 60hp Vermond et Quellennec with a rotary cultivator fitted at the rear.

Tractors started to appear in other countries prior to 1914. Munktells in Sweden made their first model in 1913, a

The first-ever tractor manufactured in Sweden - Munktell's BM 30-40. The two-cylinder engine tractor came off the production line in 1913.

very large tractor on the American pattern. Italy too made their first venture into tractors when in 1913 Pavesi introduced their Tipo B. This, like many early Continental tractors, resembled a motor vehicle of the period with its fully enclosed engine and mid-mounted bench driver's seat.

I V Mamine, in about 1911, laid the first foundations of what would become one of the largest tractor manufacturers when he built several tractors in Russia. According to A Dupouy, in his book *Les Tracteurs et Engins Speciaux Chenilles Sovietiques*, the three models were the Ouniversal of 20hp, the Prosrednik of 30hp and the 60hp Progress. These were clearly based on American machines.

Australian designs

Far away from the main centers of tractor production, Europe and the USA, four companies went into the tractor business, one of whom produced machines of a very advanced design. Australia had imported a number of tractors from both the USA and Britain and this encouraged home production. The first one was the McDonald EA of 1908 using their own two cylinder 20hp engine driving a three speed and reverse gearbox through a clutch in the pulley. Graeme Quick in his book *Australian Tractors* gives 13 as a production figure up to 1913 when an improved model, the EAA, took over until 1920. Ronaldson and Tippet built a prototype in 1910 but did not produce any more until the 1920s. The third conventional tractor was the Jelbart oil tractor using their own design two-stroke single cylinder kerosene engine of basically stationary engine style. With the improvements and a choice of engines up to 45hp they lasted until the mid 1920s. An unusual feature was belt drive with tightener in place of a clutch.

On the other hand the Caldwell-Vale was far in advance of anything else before. Descended from a large 10 disc motor plow built by the Caldwell Brothers in 1907, this tractor had equal wheel four wheel drive with brakes all round and power steering on both axles, all in 1913. With engines up to 80hp, these tractors seemed to have a very good future. Unfortunately one dissatisfied customer took the company to court and won his case. Caldwell-Vale had to sell up to pay the costs involved and so the end came in 1916. It would be many decades before a tractor like it would be seen again on Australian farms.

Pre-war

The main stream of tractor production remained in the USA during the pre-Great War period. Although the number of tractors at work in 1900 was not great, they were successful enough to encourage other entrepreneurs and some large companies to make a start with tractor production. Almost all were of the large

gasoline traction engine style destined for the wheat lands and increasingly the designs moved towards drawbar work, mainly plowing.

Several steam engine builders tested the market in 1909. Avery's Tractor Truck was a radical but potentially good idea that did not succeed, whilst Russell's built an enlarged version of the British Ivel. Buffalo-Pitts built a monster 35-70 horsepower with an unusual three cylinder engine. Case returned in 1911 with a more practical design than their earlier one and by 1913 had a reasonable model available. Also at this time the first of a long line of Avery tractors appeared with their unique system of moving the engine to change gears. This involved the complete engine and cooling system being mounted on a subframe that slid forwards to enable the driver to change the gears; the whole lot was then pulled back into the mesh. Needless to say, no one else employed this idea. Avery however, stayed

with the design to the end of their production in the 1920s.

As is so often the case, most improvements to almost anything are made by people who come into the industry without any pre-conceived ideas. Ansel S Wysong in conjunction with C M Eason took a similar view to Albone and Saunderson in Britain that the tractor should be smaller and more universal in its use. They built and successfully tested several models but commercial success eluded them. Writing in the *Scientific American* in 1907, Eason made a very prophetic remark; "Although a heavy load may be pulled on good ground, it is more effective to pull a few plows at a high rate of speed with special moldboards, if necessary, thus giving plenty of traction to cross muddy or sandy spots". It was to be a long while before this approach was put into practice.

Although they did not have the vision of Eason and Wysong, Charles Hart and Charles Parr made an impression on the United States tractor industry during its early years. They were university friends who founded a company to build stationary engines. Around 1901/2 they decided to use one of these engines in a tractor. The Canadian magazine *Nor'west Farmer* in 1929 had an article on this machine that gives an insight into the problems faced by early tractor builders. Hart and Parr used some six-inch cast iron tubes for the frame and decided to use them for the exhaust as well. However, after the tractor had run for a while they became red hot and bent under the engine's weight. This idea was quickly dropped, the frame replaced and the exhaust redirected. Then, during delivery to the buyer in Iowa, the tractor fell whilst crossing a bridge. It tumbled 16

From the stable of the highly successful American tractor builders, Charles Hart and Charles Parr: the Hart-Parr 12-27 no. 1.

feet into the water and took two days to extract. It says something for the construction that it was delivered and gave good service for 17 years.

Hart-Parr founded a line of tractors recognizable by their corrugated iron canopies and square radiators using oil as a cooling medium. The range ran from the 'relatively' small 12-27 Oil-King to a 60-100 of which only a few were made. A slightly smaller 40-80 was also made and a number were sold and used for plowing. They must have made an impressive sight when at work. Hart-Parr claimed to be the first American company to build only tractors on a production basis. The

number 13 catalog - after fully describing the 30, 40 and 60 horsepower models - shows them in use for a wide variety of field work, both in the USA, Russia and Romania.

Exports were now an accepted part of American tractor production. Kinnard-Haines was assembling Flour City tractors in Russia. Holt was also selling into Europe as were several others notably International Harvester whose smaller models were selling well. IHC also sold well in South America, Australia and elsewhere. IHC came into being in 1902 through the merging of the McCormick and Deering companies along with several

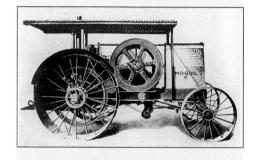

The American company, International Harvester, was formed through the merging of McCormick and Deering in 1902. They later produced two of the most famous early tractor series, the Titan and, shown here, the Mogul.

The IHC Mogul shown here with a Ransome four-bottom plow.

Above and right: The Advance-Rumely model H, one of the famous Oil Pull line which began in 1909 and lasted until the early 1930s.

others. The new company entered the tractor field by buying in chassis from the Ohio Manufacturing Company, successors to SS Morton. IHC simply fitted a single cylinder engine to the friction drive provided. This was soon modified to gear drive in order to improve the tractor for field work. Two lines of tractors came out of IHC, the Titan series sold by former Deering dealers and the Mogul series sold by former McCormick dealers.

Rumely, later Advance-Rumely, who built a range of machinery - mainly steam engines and threshing machines – brought out their first tractor in 1909. They had secured the services of John Secor who, together with William Higgins, perfected a carburettor for kerosene.

After several early models, the style settled down to what became known as the Oil Pull line which was, with improvements, to last until the sale of Rumely's to Allis Chalmers in 1931. These tractors used oil cooling with exhaust draught instead of a fan.

Crawlers, whose origins can be traced back a long way, received a new lease of life with the appearance of a gasoline half-track from Holt in 1907. Holt was a Californian manufacturer of steam engines, combines and general farm machinery. Holt's, together with their rivals Best's, were to dominate this type of tractor for much of its early development.

Another notable make was Bullock whose Creeping Grip came out in 1910. This was a full crawler with no tiller wheels. An unusual one was the Ball Tread, later the Yuba, of 1911. This used large steel balls in a race to support the track and came very close to success, lasting until the late 1920s.

By 1914 a whole range of makes and models were on sale in the USA. Many

Demonstrating its versatility, this Holt crawler of 1906, is able to mount a steep incline.

US manufacturers Holt, were to dominate the early crawler market. This immense machine was one of the first to be produced by Holt in 1904.

Herbert Saunderson spent many years working on tractor design, but it wasn't till the early years of this century that his later designs went into production. Here, the Saunderson Universal of 1908 is shown at work on the farm.

One of the early British tractors - the Petter of 1912. The Petter company had been experimenting with tractor design since 1903.

influential ideas were to be seen in some of them. Mounted plows were available on several tractors, belt pullies on almost all. High speed engines replaced what were basically converted stationary engines in the newer models. The Transit Thresher Company had, in 1906, an idea for a big tractor pulled and powered thresher that would go through the fields and thresh the grain straight from the shock. It proved impractical but gave rise to the first four cylinder, in-line engined production tractor, later called the 'BIG FOUR'. This went into the Emmerson-Brantingham Company at a later date.

Special purpose tractors included an early Holt for vineyard use, an orchard tractor from GASPORT of Gasport, New York, and a tricycle version of the Waterloo Boy with individual rear wheel turning brakes. This was the California

Special for use in orchards and vineyards. However, the implements were trailed rather than attached as on the Farmall of 10 years later. Truly a lost opportunity to develop a row-crop tractor.

The idea that all wheels on a tractor should be driven had been tried experimentally by several people. The Heer Engine Company of Portsmouth, Ohio, put theory into practice in 1912 and commercially produced one of the first four wheel drive tractors.

By the tractor's silver anniversary in 1914, all the components that make up the tractor of today had seen the light of day. The last one came in 1913 when The Wallis Tractor Company put on the market a frameless tractor, The Cub. This used a piece of boiler plate rolled into a bath shape, into and on to which the engine, transmission etc were fitted. One of the

engineers on this project was C M Eason.

In an attempt to give farmers an informed choice, several countries held official tractor trials. Canada was the first off the mark with the Winnipeg Trials of 1908. This drew six tractors, a Transit, two from IHC, and a Kinnard-Haines, all from America, and two British, a large Marshall and a Universal made by Saundersons. Sadly, the Universal broke down and was withdrawn. The Winnipeg Trials continued until 1912. Britain held tractor trials in 1910 with only Albone's Ivel and Saunderson present, and it was late in the Great War before any more trials were attempted in the UK. The USA first held trials at Freemont and Omaha, Nebraska. These spread across most of the Mid West. In the course of time, other countries tried this method of assessing tractors with varying results. The main problem was presenting the engineers' results in a form easily understood by farmers.

On 28 June 1914 in the obscure Eastern European town of Sarajevo, an assassin's bullet started a chain of events that would change the world in many ways. The tractor industry would be affected more than people expected and one tractor's influence would be fundamental to all successive machines. The Great War started exactly one month later.

Above: Darby and Maskell with their motor plow of 1913. John Maskell is seen at the controls of the machine, Albert Darby is seen in the background to the right facing the camera.
Below: An artist's drawing of the Lawter tractor brought out just before the outbreak of World War I in America.

The Great War

With the advent of war, Europe was forced to turn to its neighbor across the Atlantic to help improve its impoverished farming methods. New tractor models, such as the Fordson F, appeared in 1917, and although still novel concepts to most farmers, tractors achieved their aim as food production rose by over 50 per cent by the end of the War.

Beautifully restored - one of the highly
successful Case 10-28hp cross motor
series, brought out at the end of the War,
which really put the company on the map.

The Great War

At the onset of hostilities in 1914, people confidently expected that the war would be over by Christmas. No one it seems speculated as to which one beyond the first one. For the first time in history, aircraft laden with bombs could inflict death and injury on civilians hundreds of miles from the front line. It is interesting to note that bombing had a dramatic effect on the history of tractors, too.

As a backdrop to the next phase in tractor history, it is enlightening to examine the state of Britain and the rest of Europe at the outbreak of the War. Britain was, by 1914, the hub of a vast Empire that supplied food and raw materials on a seemingly endless scale. As a result of the cheap imported food policy, British farming was in a much depressed state. And when the German U boats began attacking the Empire's shipping, the country suffered considerably. The military authorities had systematically commandeered men, horses and enormous quantities of hay and other horse feed to replace the incredible losses at the front. Losses as high as one million horses per month were quoted. To replace these, even by importing from America, was an impossible task.

Lloyd George's coalition government of 1916 recognized that Britain would be more likely to be starved into surrender by submarine activity than militarily defeated, and belatedly tackled the farming issue. One of the first moves was the setting up of district committees to direct farmers into ways of increasing

Originally sold as the Waterloo Boy, the British knew this tractor as the Overtime. It was without doubt one of the most successful tractors of the War, being bought by various ministries of the incumbent government, among others, to help increase efficiency and economy to the farm.

production. This was mainly by plowing up the land that had gone back to grass in the previous 40 years or so. The two big questions were what to do it with, and what labor to use? The latter was partly solved by the formation of the Women's Land Army and on a small scale, when tractors were readily available, by training wounded soldiers who were unfit for further military service.

Finding the tractors was a different job.

By 1916, of the pre-war makes, only Saunderson tractors were still in production. Rustons, who had made a few Hart-Parr 30-60 Old Reliables under licence, had, like Marshall and the others, gone over to war production. Some people had seen the forthcoming need for

tractors, and in 1915 the Weeks-Dungey and several motor plows entered small scale production. They were, however, a drop in the ocean.

As home production was negligible, so the answer was to import from the United States. A ship load of tractors would have more effect on increasing food stocks than the same ship load of food. This resulted in all sorts of tractors being brought in, regardless of their suitability or reliability, and a lot were questionable on both counts. It also gave rise to a degree of deception as importers sought to imply that their tractors were British by renaming them.

The Austin Motor Company offered the Culti-tractor Model 1, this was a Peoria,

Above and right: IHC's ever-popular Titan. This US machine and its sister, the Mogul, was well-liked and respected during the long war years.

from Peoria, Illinois. The almost identical Big Bull tractor sold as the Whiting-Bull, and the Parrett was called the Clydesdale with a heavy horse in the background of the company's tractor advertisement. The most famous was the Waterloo Boy known in Britain as the Overtime, and sold by the Overtime Farm Tractor Company of the Minories, London. Around 4,000 examples of the Overtime were purchased. Their advertisement lists as customers His Majesty's War Office, the Ministries of

Munitions and Agriculture, the Food Production Department, Department of Agriculture for Ireland, numerous War Agricultural Committees and the Army Canteen Committee as customers. Quite what the latter used them for is not stated. At £325 it brought efficiency and economy to the farm. And in a rather strange statement it says "there are more Overtime tractors at work on farms in this country than those of any other three makes put together". Given that, for

example, only about 10 Allis-Chalmers 10-18s and probably fewer Rumely 8-16 Motor Plows and other odd makes came over, this was an unnecessary claim, especially in view of the Overtime's obvious success.

IHC, who had branches in Britain for quite a while, wisely sold the two models most suited to British conditions and backed them up with service and spares. The Titan 10-20 and Mogul 8-16 were to make IHC well known and liked; in fact

Harry Ferguson

Among the Inspectors of Tractors working for the Government in Ireland, at that time all still part of the United Kingdom, was Harry Ferguson. Harry, a farmer's son with mechanical talent, had left farming and gone into the motor trade at his brother's garage in Belfast. By 1917 he was working for the Irish

Board of Agriculture and his experience with the available tractors convinced him that there had to be a better way of coupling the tractor and implement than by drawbar and pin. Briefly popular at this period were kits to convert cars, mostly Ford Ts into tractors. Harry built a two-bottom plow directly attached to the back of one of these T tractors. Raised and lowered

into work by a lever, it did away with wheels, drawbar and quite a bit of the usual plow framework. The industry would be hearing a lot more of Harry Ferguson after the war.

Above: Harry Ferguson was quick to capitalize on the opportunity to convert cars to tractors. His two-bottom plow attached to a Ford model T was to mark the beginnings of an immensely successful career in tractor design.

3,500 Titans and a smaller number of Moguls came over by 1920.

Among the tractors that came into Britain were a number of crawlers, some like the single-track Bates Steel Mule were next to useless, others had a lasting influence. The Cletrac was an excellent tractor, sold well, and remained a popular make until the end of the Cletrac Company in the 1940s. Holt, besides selling a few of their smaller models for agricultural use, supplied a large quantity of 75hp half-tracks to the army. To meet the demand, a number were made under licence by Ruston's of Lincoln and used by both the British and Russian armies. A three track Killan-Strait, powered by a four cylinder 20-40hp Waukesha engine, was one of a number of machines tested by a committee for the British army as part of their Landship Project. One feature that went into tank design was the sloping track that enabled tanks to climb over obstacles. The two driven tracks of a Killan-Strait had the slope at the rear to enable it to back out if it became bogged down. However, one name overshadowed all these; Fordson.

An early Ford prototype

Fordson

Henry Ford had been experimenting with tractors since at least 1907. He wanted to do for farmers what he had done for the motoring public with the Model T car and give them a cheap, reliable tractor. By 1916, he had several revolutionary prototypes at work. These used a three-piece cast iron unit instead of the angle or channel iron frames in general use. Ford had set up a separate company to pursue his tractor ideas and among the staff were 'Cast Iron' Charlie Sorenson, Joe Galamb and Eugene Farkas, who

Left: The controls layout of the Fordson model F.

Below: The famous Fordson model F which first went into production in 1917.

between them came up with the basis of what later became the famous Fordson F tractor. With its stressed cast iron construction that contained all the working parts in dust-proof and oil-tight units, it got rid of most of the early tractors' weaknesses. In the fullness of time practically all the world's tractors followed this pattern of cast frame.

However, before Ford had a chance to put the F into production, his pilot models in Britain – introduced at trials for a Plowing-up campaign – were hailed a success, and he was requested by the British Government to start production in Britain. Sadly, on the 13th June 1917, seventeen Gotha bombers of the German command bombed London in daylight. The government immediately ordered all available production to be turned over to fighter aircraft manufacture to defend Britain against this new menace. Ford lost his proposed factory and was asked if he would make the tractors in the USA

instead. This was agreed and by October 1917, an almost unbelievable four months after the setback, Fs were starting to come off the line for shipment to Britain.

This was not the only setback for Ford's tractor. The widespread news of his interests and intentions had been noted by several speculators who were hoping to cash in on Ford's reputation. In Minneapolis, rivals formed a Ford Tractor Company using the name of one of their engineers, Paul Ford. But the plans went wrong, the tractor was no good and, even worse, they sold one to a farmer who was also a member of the Nebraska State Legislature. His experience with this and a Bull tractor led eventually to the Nebraska Tractor Tests which later achieved worldwide acceptance. Deprived of the right to use the name Ford on his tractors, Henry used one based on his tractor company's name, Henry Ford & Son, hence the tractors were called Fordsons.

The French farming scene was possibly even more behind than that in depressed Britain. Many tractors were American machines renamed. This Tourand-Latil is an exception.

Food production up

Despite all the problems that arose through the use of what was to most farmers a novel machine, tractors achieved their aim. In 1918, the wheat harvest was up by over 50 per cent on 1916, barley and oats also rose, and potato production was up by almost 70 per cent.

From the English Channel to the Swiss border, a swath of terrible destruction spread across Europe's farmland. In some areas it would never be reclaimed, such was the devastation.

The situation in France was very similar to Britain; depressed agriculture, which in some areas was also rather backward, supported practically no home tractor industry. Importation was again the answer and American tractors helped French farmers to meet the wartime food production programme.

The French also indulged in the practice of renaming American tractors. Allis-Chalmers 10-18s became the Globe and the Bean Tracpull 6-10 was called the Csar, both sold by a company in Paris. The Galloway Farmobile 12-20 was sold as the Le Gaulois tractor. Galloway seemed to have supplied this model only to France because the Galloway sold in Britain by Henry Garner as the Garner was the Galloway Bear-Cat 10-20. Garner's claimed to have built this tractor; more likely they assembled it. The Bear-Cat also appears to have been sold as the Maxim in the United States. Galloway's went out of business when one of their clients failed to pay in full.

Experimental work went on in Italy and at the War's end two good tractors went into production. They were the Fiat 702 and the Pavesi four-wheel drive, articulated steering tractor. This was the first successful articulated tractor.

In neutral Sweden, a second company joined Munktells in tractor production. Avance in 1917 offered the first semi-diesel engined tractor. In Sweden, as in most European countries, oil products had to be imported and consequently the ability to use as cheap a fuel as possible and at least in the early part of the century, a range fuels, led to the adoption of the semi-diesel engine. This engine, using ignition by a red hot bulb in the cylinder head and injected fuel, possessed the necessary characteristics to accomplish both objectives. The tractor was a mix of old, i.e. chassis and tank cooling, and new, the engine and a power lift for a three-bottom plow.

On the other side of the Western Front, Germany and its allies were also suffering food shortages as a result of the Allied naval blockade. The blockade runners that did get through were more likely to carry

The Emmerson-Brantingham with its fully enclosed engine.

The massive Holt 75 half-track was widely sold in Germany and Eastern Europe before the War.

war materials than food. Although in theory they too could have imported from the United States, neutral until 1917, in practice this did not happen. Even under these circumstances, and with huge losses of men and horses like the Allies, German tractor production was not very great.

Motor plow production by the pre-war makers continued, Stock being one of the larger producers. More conventional style tractors came Lanz, Stoewer and Hansa-

Lloyd among others. These were large tractors more suited to the estates of the German landowners than the average farmer. The Lanz Landbaumotors powered by four or six cylinder engines of up to 90hp usually had a power driven rotary cultivator at the rear. Stoewer offered a choice of the 3S17 with a 40hp four cylinder engine or the 70 hp four cylinder 6S17. These had a fully enclosed engine rather like the Emmerson-Brantingham

Abandoned is this 1916 Avery 8-16hp tractor: the company's answer to the need for a smaller tractor.

series, with a canopy for the driver. They also had provision for a power lift for an integral plow. A smart cab was available for the Hansa-Lloyd HL18 built in Bremen.

The Germans, like the British, leaned more to automobile practice for both style and components than did the American makers, many of whom were still using such primitive ideas as fifth wheel steering. Holt's had sold quite a number of half-tracks, mainly the 75hp model, into Germany and Eastern Europe and the Kaiser's army, like the Allies, used them as gun tractors. German gun tractors were more often like very large agricultural tractors than the vehicle based ones used by the Allies, for example the American

FWD. After the war, some ex-German gun tractors had a second lease of life in the timber industry.

"American citizens are at liberty to ship all articles whatsoever to the nations engaged in war" stated the State Department in Washington. Despite any moral misgivings, this was an industrial opportunity as never before and it was used by the tractor manufacturers as well as others. The slaughter of horses in the fighting created an export trade in American horses and mules. This in turn led to a demand for replacements, which is where tractors came into their own. Europe also needed food and by buying a tractor the land released from feeding

Cropmaker 17-27
1917

Hart-Parr
"27"

A promotional drawing of the Hart-Parr Cropmaker 17-27, built in 1917.

horses could grow saleable crops. To the smaller farmer, this was an attractive prospect, so the demand arose for smaller tractors. Prior to 1914, most American tractors had been too large for many farms so in effect the flood gates opened to would be tractor makers, and they poured out. A large book could be written on American tractors of this era.

First threat to the established makers came from the Bull tractor manufactured by Minneapolis Steel and Machinery for the Bull Tractor Company. A very basic tractor on a triangular frame, it used a single drive wheel to eliminate the differential, one wheel at the front steered it and the third simply stopped it falling over. A two cylinder opposed

engine of 5-12hp and a simple forward and reverse transmission completed it. The selling point was the price – US$335 FoB. It sold like the proverbial hot cakes to farmers eager to mechanize their land work. With so many at work, faults soon showed up and sales dropped as rapidly as they had risen. An uprated model was offered from 1915 without success and the company faded away. The Bull did however alert the industry to the vast market awaiting a successful small tractor. Avery's answer to this was to shrink their existing design down to an 8-16hp replica of the largest 40-80 tractor in their line. Following the Fordson's announcement they offered a three wheel, 5-10hp with a four cylinder in-line

engine. Apart from the later Track Runner, this was Avery's only attempt at a conventional tractor.

The other major brands such as Hart-Parr, Twin-City and Minneapolis took a similar line. Rumely tried a motor plow and Case, who had hardly been involved with large tractors, offered a three wheeler – the 10-20, followed by the four wheel 9-18 in 1916. This set the style for the very successful cross motor series that really put Case on the tractor map.

It was among the small companies that innovations took place, some wildly optimistic and some very advanced. Most failed through lack of finance which resulted in low volume and high selling price or bad management, and lack of

service and spares. Odd ones included the Ebert-Duryea with its gearless drive; conical rollers ran tightly between two rims on the drive wheels. The Detroit used reins to operate it from the implement.

This idea for driving appears in many countries at various times. The Lang with an unusual three wheel layout of two front drive wheels and the rear wheel on a hinged frame offered power steering. The Fagiol of conventional style used a tiller instead of a steering wheel. There was no agreement as to what a tractor should look like. Three wheels were a popular idea. Allis-Chalmers favored the front steering wheel in line with the right rear wheel, while the Happy Farmer thought the steered wheel should be way out in

At the outbreak of the War, IHC found a welcome market for one of its most successful models, the Mogul. This one dates to 1914.

53

The Great War

Manufactured by Saunderson & Mills Ltd in Bedford, England, this Saunderson Model G was used as a landing ground obstruction during World War II. It was built between 1910 and 1916. This model gained a gold medal at the Dartford Plowing Match in 1913.

front on what was almost an extension of the chassis. Kardell decided the steered wheel should be at the rear. How many wheels should be driven? Michigan decided on all three, the front steered wheel being driven from above and capable of being turned through a full 180 degrees. On the question of the number of driven wheels, some manufacturers– including Utility – decided to drive all four. Most settled for

two, but quite a number of single wheel drive models were made.

Possibly the stangest tractor of all was the Chalmers. Perhaps the designer lost his nerve half way through because to a fairly conventional rear end and engine he added a pole hitch for two horses instead of a front axle. The driver sat in front of the truncated tractor and with reins in hand and presumably a foot on the clutch, drove the whole combination to the best

of his ability. It has been described as an outboard motor for farm horses which seemed to fit it exactly. R B Gray's book *Development of the Agricultural Tractor in the United States* devotes one whole page to diagrams of layouts, 27 in all.

Crawlers were the subject of a lot of layout experimenting. The Killan-Strait, mentioned earlier, with two rear driven tracks and a short front one for steering, was a variation on the tiller wheel steering still favored by Holt and Best. The two front wheels and a single rear track used by the Bates Steel Mule and Tom Thumb tractors were an attempt to make the crawler suitable for row-crop cultivation. Monarch, Austin, Bullock, Cletrac and the smaller models from Holt and Best used the standard two track layout for drive and steering. that would cultivate between the rows to cope with corn, a major crop in the US. This needed quite a high clearance plus maneuverability, things not readily available on the Great War era tractors. The initial solution was a motor cultivator. This led eventually to the Farmall style row-crop tractor. The majority of early cultivators were too underpowered and ungainly for anything other than their specifically designed purpose and, like the motor plows, soon faded into insignificance when row-crop tractors went on sale.

Motor plows enjoyed a brief surge of popularity with varying layouts and degrees of success. The Square Turn with friction drive to each drive wheel, could do just what its name implied by reversing the drive to one wheel. With the plow removed and the seat reversed it could be used as a tractor and steered by what was

One of the most popular motor plows, the Moline Universal One-man tractor shown in use for an agent's promotion. This model dates to 1917.

now its front wheel. After initial success it went out with the other motor plows in the early 1920s. The Moline Universal was probably the best of them and put two new items into production. The electric starter motor and lighting had first been used on the prototype Lord tractor in 1914. Moline made the first production use of these features. The Moline was also widely exported, not only to Europe, including Britain, but also to Australasia.

The V8 engine made its debut in the Common Sense tractor. Change on the move transmissions were offered by Heider and by the John Deere Dain. The Dain with its one rear and two front steering wheels, all driven, was before its time, and Deere bought the very popular Waterloo Boy Tractor Company. They were the descendants of the Froelich and since 1912 had made a two cylinder tractor of

about 20hp which they had improved until in pre-Fordson days, it was one of the best available in its class.

Wallis offered a smaller version of its Cub, the Cub Junior, also a three wheeler. This really introduced the concept of everything being enclosed, in contrast to the spread-out layout of some competitive tractors.

Plowing had been a major tractor job almost from the beginning and in the USA at least the plow was a large implement with a wooden platform on top for the plowman. His job was to make adjustments while at work and lift the bodies out at the end of the furrow, and drop them in again after the tractor and plow had turned around. In Britain and Europe the small plows had the plowman sat on the plow with the levers nearby. In about 1915 the invention of the self-lift

The Czech Oekonom tractor of 1919 from the manufacturers Tovarna Motorovych Pluhu. This model bore a four-cylinder engine and a five-speed gearbox.

plow did away with the plowman and tractor plowing became a one-man job. Another solution that occupied a lot of people, both in the USA and Europe, was a directly attached plow lifted by engine power through a winch or lever and cam system among others. Emmerson-Brantingham persevered with the directly attached plow and came very near to success with their chain lift.

There had never been anything in history like the Great War and when it ended in November 1918 the mechanical legacy it left was considerable. The aircraft industry, arising from almost nothing, had produced high speed, high power to weight ratio engines. Also from this industry came rubber tires which would withstand the stresses and strains of landing large aircraft. The vast output of vehicles of all kinds had brought transmissions and associated parts like the clutch to a high state of efficiency and reliability: unreliable military equipment does not result in repeat orders to the manufacturer.

The tractor which had contributed to the War Effort, could now draw something back. Many of the service men returning home were familiar with machinery and its maintenance: those who went into farming would now be better equipped to use to the full the next generation of tractors.

Manufactured by the Moline Plow Company of Moline, USA circa 1917. This impressive machine housed a two-cylinder horizontal opposed engine rated at 10-12hp. It had one forward and one reverse gear and was imported into Britain by British Empire Motors of London.

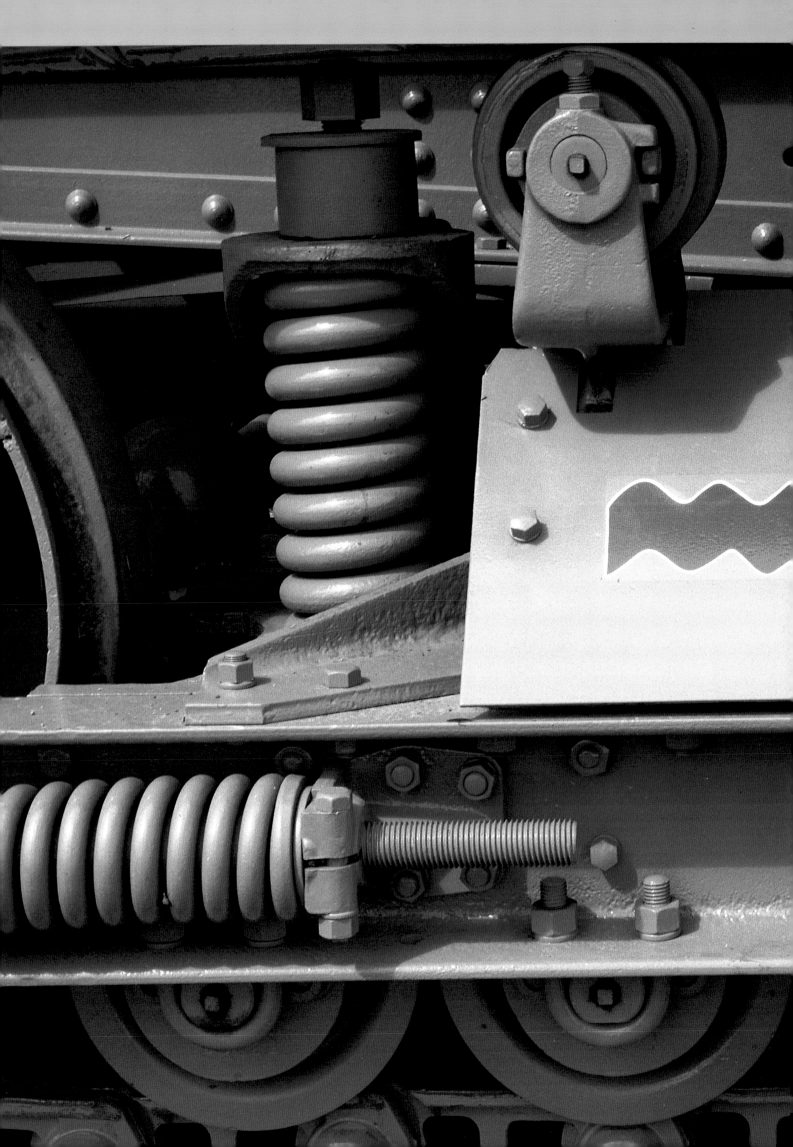

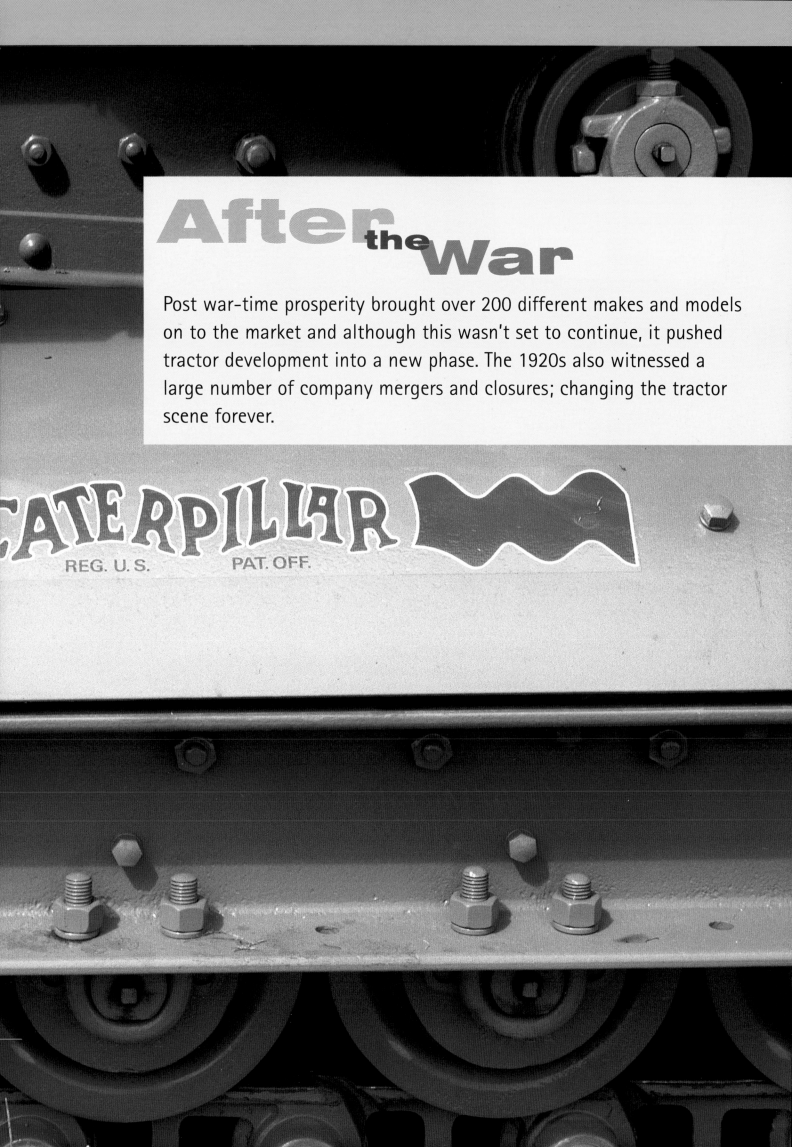

After the War

Post war-time prosperity brought over 200 different makes and models on to the market and although this wasn't set to continue, it pushed tractor development into a new phase. The 1920s also witnessed a large number of company mergers and closures; changing the tractor scene forever.

The United tractor produced in 1929 by the United Tractor and Equipment Corporation of Chicago, alias Allis-Chalmers, who were one of 32 companies involved in the corporation. The tractor was so successful, it survived the corporation's swift demise and became part of Allis-Chalmer's own line for almost 20 years.

After the War

The first year of peace saw an unprecedented output of tractors. The American farmer discovered he could choose a tractor using any criteria he liked. Horse power ran from 1½ for the newly introduced Beeman two-wheel garden tractor to the giant 120hp Holt half-track, probably ex-army. Literature of the period lists tractors from the ACME in four-wheel or half-track form to the very odd Zelle. The Zelle was, to say the least, unusual even compared to its contemporaries. It appears to be an attempt at a high clearance tractor because the engine and most of the frame were above the front wheels and the axle of the rear wheels. The driver sat on top of all this looking down on literally everything, tractor, implement and crop!

In between these two, almost 200 makes and models were offered. Quite how many reached production, let alone sales, is a different matter. Some probably did not even get beyond a prototype. Charles Wendel lists quite a number of companies in his *Encyclopedia of American Tractors*, on which no information is available.

The slight of hand previously used with names in Britain was nothing compared to the fraud by some American makers. A company called the Pan Motor Company of St Cloud, Minnesota, offered in its advertisement a small half-track with the tracks at the front and a hinged two-wheel unit behind with a seat. Called the Tank-Tread it was claimed to be the farmer's war tank. The sole prototype only just managed to run under its own power and the company promoter Mr S C Pandolfo was later found guilty of fraud.

The post-wartime prosperity looked set to last forever, and sales continued apace for a while with the Fordson taking an

A 1924 model of the Fordson F. At the start of the 1920s, Ford accounted for almost 75 per cent of tractor sales: competitors retaliated by offering a host of extras on their machines.

Holt's 2-ton crawler with a potato digger at work in California in 1924.

increasing amount - almost 75 per cent of sales - at one point. After fulfilling the British wartime order, Henry Ford had released his tractor on to the American market where, in a short time, it cut a swath through the domestic industry. Very basic, it cost little to produce compared to many others, and sold for a lot less.

The Fordson basic specification spawned a host of companies offering extras; governors, alternative ignition systems, half-tracks and full tracks, to name but a few. Fordsons turned up everywhere, and tractor makers worldwide felt the threat of Henry Ford's production constantly behind them. For many, as in the United States, it proved too great a worry to bear and they gave up.

However, not everyone collapsed in the face of the Fordson. IHC, who had gone

through the war with basically their pre-war line, were knocked into second place by the Fordson. But they bounced back with the introduction of the 8–16 Junior. This machine had the option of a power take–off, mainly with a view to the binder market, for which they offered a p.t.o. driven model.

Writing as "The Reflector" in *Implement & Tractor Magazine*, Elmer J Baker tells a story of a Junior tractor and p.t.o. combine being demonstrated in an American rice field. The combination did everything that was expected of it, however the demonstration had been watched by the local parish priest who then spoke to the rice grower. When the grower declined to buy a tractor and combine, he explained to the salesman that the priest had asked him to consider

what would become of all the farm hands who would lose their jobs. There would come a time when replacing farm workers would be an important issue.

Market collapse and changes

The post-war prosperity bubble burst in 1920, and farm incomes started falling. Tractor company doors closed as fast as they had opened only a few years before. To counter Ford's price cutting, IHC offered a free plow with each Titan, now a very dated design. IHC cleared out its entire tractor line and, with the introduction of the 15-30 in 1921, and the 10-20 in 1923, gave Ford some real competition for the first time.

These two IHC models with their Ford-style cast frame incorporated all the

things the Fordson lacked: magneto ignition, a better clutch and rear-axle drive plus a built-in p.t.o. They set a new standard for tractors. IHC topped this in 1924 with the introduction of the Farmall, the first real row-crop tractor. This would not only do the plowing and other land work, but when fitted with one or more cultivators mounted either in front, under the middle or at the rear, would work along the rows of corn, cotton and other row crops. To a lot of American farmers this was just what they had been waiting for. Although Farmalls were exported in some numbers, and years later IHC factories in Europe and elsewhere would make their own versions, the American row-crop style tractor never gained the popularity it enjoyed in the land of its birth. Different cropping methods and soil

The IHC 8-16 Junior model which helped the company stem Ford's intense competition. The machine had the option of a power take-off.

types did not favor the V–twin front wheel layout that was popular in the United States.

Case's interpretation of the cast-frame used a cross engine for the smaller models of a series ranging from the 12–20 to the large 40-70 hp. The larger ones still used a channel steel frame. A very successful and popular line, they lasted until 1930 when Case followed convention and used in-line engines.

The second interpretation of the cast-frame came from John Deere. The long–running Waterloo Boy had, by the early 1920s, become outdated in spite of

detailed improvements, the last one being the adoption of Ackerman steering. The famous D went on sale in 1924. Initially like the Fordson, it was very basic with its two–cylinder horizontal kerosene engine and a two speed and reverse gearbox. Yet it was to remain in production until 1953. During its life it underwent many changes while retaining the basic 1924 concept.

To meet the Farmall competition, Deere brought out a small two–cylinder row–crop tractor in 1928 – the GP or General Purpose. This brought with it a power lift for the tool bars driven off the transmission. Advertised as the tractor

The cross-motor Case tractors were a very popular line in the 1920s.

A 1924 John Deere model D. It housed a two-cylinder horizontal kerosene engine and was to remain in production - with many improvements - until 1953. The first fifty D models displayed a flimsier 'ladder-style' front axle unlike that shown here.

The John Deere General Purpose or GP - a two-cylinder row-crop tractor brought out in 1928. It was promoted as the tractor with four power outlets: drawbar, belt pulley, power take-off and power lift.

with four power outlets: drawbar, belt pulley, power take-off and power lift, Deere also stressed the simplicity of the two-cylinder engine in all their sales literature, something they did right to the end of two-cylinder production in 1960.

Possibly the best Fordson copies were the Twin City 12–20 and 20–35 from the Minneapolis Steel and Machinery Company. These were very well engineered machines using valve in head engines with four valves per cylinder. They replaced the earlier prairie-type tractors.

Allis–Chalmers joined the now conventional style of machine with their 18–30 model. Originally, in 1914, Allis had offered a licensed version of a Swiss tractor, the Meyenburg, one of the many European tractors with a power-driven rotary cultivator fitted at the rear. This did not appeal to the American farmer and A–C then produced the 10–18 three-wheeler during the Great War.

After making some of the largest tractors, and in large numbers, Hart-Parr

redesigned their tractors for the expected post-war boom. The first was the new Hart-Parr 12-25. With its two-cylinder horizontal engine it set the pattern for the rest of the line. One feature offered later on was a p.t.o. driven off the crankshaft counterweight via a separate clutch and capable of transmitting the full power of the engine. This in effect gave a direct engine driven p.t.o, the first since the original Gougis. However, this was never really promoted in the sales literature and the opportunity to steal a march on their competitors was lost.

The Hart-Parr 12-24E: one of the first of the smaller machines built for the post-war boom.

The Moline Motor Plow model D - a redesign from the older Universal and built between 1918 and 1923. It had a four-cylinder engine, electric starter and was rated at 9-18hp. Ballast weights were set inside the drive wheels to lower the centre of gravity.

General Motors

A newcomer to the industry was General Motors. The Model T Ford car had taken a large slice of the popular motoring market and the Fordson tractor had done the same with the small tractor market. Stung by the loss of sales, GM's chairman William Durant, decided to challenge Henry Ford on his own ground. With no tractor experience, GM looked round for a takeover and found one in the Samson Sieve-Grip, a moderately successful tractor from California. Modified and transferred to a new factory at Janesville, Wisconsin, it proved to be too expensive to sell against the Fordson and was replaced by the Samson M. This was a Fordson lookalike with a gap between the radiator top-tank and the fuel tank. A promising tractor, it ceased production when GM reversed their policy on tractors after buying another company to offer a motor-cultivator alongside it. The rein-operated Iron Horse with its four-wheel drive and skid steering was as bad as the M was good and GM cut their losses and went out of tractor production altogether.

In 1919 the three-wheeler Wallis Cub Junior was replaced by a four-wheeled version, the K. Progressively improved as the OK and again in 1927 as the 20–30, the Wallis was a very popular tractor. In 1928, Massey Harris from Toronto, Canada - who had sold other makes in an attempt to establish a footing in the tractor market - bought out the manufacturers of the Wallis, the J I Case Plow Works of Racine, Wisconsin. Case Plow were distant relatives of the other Case company and Massey Harris sold the Case Plow name to them, thus ending any confusion. Massey expanded the line in 1929 with the 12–20 built to the same boiler plate pattern as the 20–30. These set the style of Massey Harris tractors for the next decade.

A British Wallis of 1920. This machine was built by Ruston and Hornsby Ltd of Lincoln, England. This design was based on the Wallis of the J I Case Plow Works of America. When the Wallis tractor of America was taken over by Massey Harris in 1928, the British Wallis ceased production.

Mergers and closures

Many of the manufacturers of large prairie tractors were in financial trouble and there were many mergers and closures. Case bought the Emmerson-Brantingham company and Aultman & Taylor went into Advance-Rumely. Aultman-Taylor had built tractors since 1914 and had earned a sound reputation. Advance-Rumely continued to improve the Oil-Pull line with a pressed steel frame and even a p.t.o. driven off the belt pulley. However, the basic design was very dated and totally unsuitable for row-crop use. In the late 1920s Rumely took a big step and introduced the Do-All row-crop tractor with a conventional four-cylinder engine and, shortly before being taken over by Allis-Chalmers in 1931, the Rumely 6. This six-cylinder tractor represented a dramatic change in style rather reminiscent of what John Deere would do 30 years later.

Avery went bankrupt and was reorganized, and Kinnard-Haines, ceased production and was eventually sold.

The Advance-Rumely Oil Pull - a successful machine in the past - soon became outdated in the 1920s. This model H from 1920 was succeeded by the light weight series from 1924.

Made in Ohio, USA, the Huber Light Four housed a Waukesha four-cylinder 12-25hp engine and two forward gears. Manufactured between 1917 and 1928 it cost over $1,000 to buy.

Most small makes disappeared one by one, but not all: some who were making good tractors survived, among them Huber, another old make, and Eagle, one of the last companies to make old-style tractors. Baker took a slightly different approach and bought in components to assemble excellent tractors that remained in production until the beginning of World War II.

Mid-way through the decade a momentous merger took place. Holt and Best, rival crawler makers, agreed to join forces. The Caterpillar Tractor Company, as the new company was called, proved to be one of the most successful tractor makers in the world, so much so that their name became synonymous with almost anything on tracks. The Caterpillar name came from Holt, who were building full crawlers in 5 and 10 ton sizes that had found

increasing use in the construction industry. Best contributed the Thirty and the Sixty; these were so popular they lasted until 1931.

Competition to Caterpillar was in reality very limited and only two makes really offered any. Monarch had only built crawlers since their formation in 1913, and by 1928 when Allis–Chalmers bought them out to get into the crawler market, they were making good tractors for the top end of the market. The other manufacturer, Cletrac, leaned more to farm use where they were very popular for general work.

A potential alternative to crawlers, the four–wheel drive tractor, had in the immediate post-war years produced a number of promising designs. Cost and the added complication of the driven and steered front axle prevented them from

The immense Caterpillar Sixty: a popular machine that came from the stable of manufacturers Best who merged with Holt in the mid-1920s.

Nebraska tests

All this activity took place against the backdrop of the Nebraska Tests. In an attempt to prevent exorbitant claims by tractor makers, the law passed in 1919 and put into operation in 1920, provided for a series of tests on tractors. These examined horsepower, fuel consumption and engine efficiency, plus a selection of drawbar and beltwork tests. The Tests were carried out at the Nebraska State University, Lincoln, Nebraska. An important requirement was that tractor manufacturers must print all or none of the results; this stopped any sharp practice in advertising by leaving out criticisms.

Early test results showed up variations in engine output. Three of the early tractors tested, the Waterloo Boy N 12-25 and the Case 10-18 gave almost exactly their makers' ratings, as did the Oil-Pull H 16-30. The Aultman & Taylor 30-60 put out 58 drawbar hp and 75 belt hp, well in excess of its ratings. On the other hand, the Avery 40-80, whilst producing just under 50 dbhp, only managed 69 at the belt. Avery then re-rated it as the 45-65 for future sales. The friction drive Heider C 12-20 from Rock Island Plow, Illinois, was the first one tested with more than two forward speeds, it had eight in all. The fairness and thoroughness of the Nebraska Tests led to their acceptance not only all over the United States but across the world as well.

becoming popular. Despite this, around a dozen companies produced four-wheel drives, if only for a short period. Wizard, who used skid steer, and Topp-Stewart who offered a road haulage version with solid rubber tires, lasted into the late 1920s. The Fitch FOURDRIVE first built during the Great War remained on sale into the early thirties and was the only one to be exported. A precursor of the future was the Roger's articulated tractor with a hinged frame that allowed both steering and the rear axle to follow uneven ground. Rated at 85hp, this was the first large tractor of the type that would become popular about 50 years later. An unusual feature that it shared with a contemporary - the Nelson four-wheel drive - was that the driver sat sideways so he could see behind as well as in front without having to turn round. Rogers had industrial use in mind more than agriculture with this arrangement. This tractor also had the distinction of being the first four-wheel drive to be Nebraska Tested.

Henry Ford who had contributed so much to the farm tractor, announced in

Above: A 1929 Allis Chalmers United.
Below: Two model N Fordsons at work. With the cessation of tractor production at Dearborn in 1928, manufacture eventually moved to Dagenham in England in the early part of the 1930s.

Above: The popular Fordson model N.
Right: A post-war Moline dating to 1919.

1928 that he was to cease producing tractors at Dearborn. The old Cork, Ireland, Factory where Fs were built from 1919 to 1923 was reopened for a while, building the improved N until production moved to Dagenham, England, in late 1932.

1929 saw three major reorganizations among American farm equipment makers. The largest and shortest lived was the United Tractor and Equipment Corporation with headquarters in Chicago. This vast operation of 32 companies included Allis-Chalmers who were given the job of providing a tractor. The United, as it was appropriately named, survived the corporation's collapse and A-C renamed it the U in their own tractor line. With successive improvements it ran for over 20 years.

Minneapolis-Moline brought together the Twin-City tractors from Minnesota Steel and Machinery, the Minneapolis tractors from Minneapolis Threshing Machine Company and the Moline Implement Company.

Hart-Parr, Nichols & Shepard and American Seeding Machine Company joined up with Oliver Chilled Plow of South Bend, Indiana, to form the Oliver Farm Equipment Company. The new company cleared out all tractor models current at the merger ready for a completely new line.

Europe after the war

The main preoccupation in Europe following the Armistice in November 1918 was to clear up the ruined countryside along the Western Front. The dissolution of the Austro-Hungarian Empire and the emergence of the Soviet Union following the Russian Revolution redrew many frontiers. All these would have a bearing on the future European tractor industry.

Britain was virtually undamaged but had to readjust itself to peace-time conditions. With the lessons of the war and the availability of ex-munition factories, it looked like a good opportunity to turn swords into plow shares, or better still, tractors.

The DL Motor Company took over one such factory at Cardonald near Glasgow, Scotland and built the Glasgow tractor from 1919. A Fordson-style cast-frame was carried on three wheels, all driven. The single rear wheel needed no differential and the two front wheels used ratchets in the hubs to eliminate the front one. Around 400 were built before production ended in 1924.

The Austin Motor Company, who had produced vehicles for the War Effort and sold imported American tractors, modified one of their 20hp engines and offered a

The 30hp Peterbro made in England by the Peter Brotherhood Company. Its unusual engine was based on a tank engine and aimed to eliminate the problem of sump-oil dilution. However, it was expensive to make and the design limped along only for about 10 years.

Fordson lookalike in 1919. This sold reasonably well despite being dearer than the Fordson F now coming from Cork, Ireland. The Austin went on until the mid-1920s when production moved to France.

The 30hp Peterbro made by the Peter Brotherhood Company of Peterborough, England, used a very unusual engine. This engine - which had its origins in a tank engine - was designed by Harry Ricardo who hoped to eliminate the problem of sump-oil dilution common to low-compression kerosene engines. He achieved this by using a two-part piston which had compression rings on the top

part and oil control rings on the bottom part. In between was a gap which coincided with a ring of tiny holes in the cylinder walls. The theory was that any fuel that leaked down and any oil that came up could be bled off. A complex and expensive engine to manufacture, it was not helped by being fitted to an old-fashioned channel frame and transmission. Never made in large numbers, it nevertheless struggled on for about 10 years.

An equally unusual war-time engine from a searchlight generator provided the basis for another tractor, the Blackstone

1930 Renault advertisement for their RK tractor, with its rear radiator and air cleaner at the front, in the usual radiator position.

from Stamford, England. Sold in both tracked and wheeled versions, its three-cylinder kerosene engine used fuel injection and a compressed air starting system which did away with gasoline starting and warm-up period before switching over to kerosene, the method common to all tractors sold in Britain at this time.

Two British companies built tractors based on American designs. Ruston's of Lincoln took out a licence to build the Wallis. Vickers of Newcastle-Upon-Tyne had an eye to the export market, so built the Aussi, closely based on the International 15-30. Neither were produced in large numbers.

Of the other attempts to manufacture a British tractor, only the Clayton crawler was produced in any quantity. The Clayton, with a Dorman four-cylinder kerosene engine, was first produced in 1916 and as well as agricultural work found employment with the Royal Air Force for towing aircraft.

All British tractors suffered from the Fordson competition of low price and good sales and service. By the time Ford ceased production in February 1928 it was too late for the British companies to recover. The generally depressed state of farming following the collapse of the post-war boom added the final nail to their coffins. By 1929, with exception of the Fordson N from Cork in what was now the Irish Free State, tractor production had effectively ceased in Britain. With the old makers out of the running it was now left to newcomers to carry on the British tractor industry.

France

The French government in a major effort to help the post-war recovery made 100 million francs' worth of interest-free credit available to the farming industry. Farmers could also get priority for any fuel they required. Tractor trials were held with both imported and French makes starting at Rocquencourt in March 1920. In October, the Chartres Trials attracted 116 models from 46 European and American manufacturers. American tractors included Case cross motors, Internationals and the Grey Drum Drive which had been one of the early imports during the Great War. The Fordson also

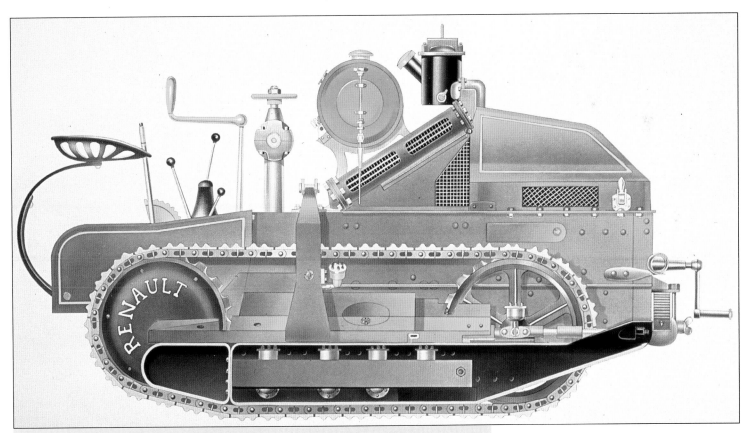

With the signing of Armistice in 1918, tank production came to an abrupt halt in France. Seeing the devastation of the French countryside decided Louis Renault to turn his attentions to agriculture. The result was the introduction of the HI type tractor - a crawler with a 20hp gasoline engine, a cone clutch, and three forward and one reverse gears.

made its first French appearance at these early trials.

Renault, who had made light tanks for the French Army, produced the HI crawler using quite a lot of tank parts. This was an odd looking machine powered by a four-cylinder 20hp gasoline engine. The radiator was behind the engine on a slope, with the fuel tank across it. The hood was the same style as the 'coal scuttle' type used on Renault vehicles. Later the HO wheeled model joined the crawler.

Another vehicle maker, Peugeot, also offered a crawler, the T3, of similar size with differential steering brakes operated by a steering wheel.

The third French crawler of the early 1920s also came from a vehicle builder, Citroën. This was a neat and narrow model with a view to vineyard use as was the wheeled version. The wheeled tractor had independent rear wheel brakes and used a Citroën car engine. Neither was in production for long however. The next tractor that Citroën sold was the car-based half-track model using Kegresse track. This was a wire and rubber based track rather than the usual metal pad and link chain used on full crawlers. The Citroën-Kegresse remained in production until World War II for agricultural, industrial and military use. The Somua, with rear power-driven rotary cultivator, continued in production as did the Amiot, one of the early makes with an integral plow. This plow was a three-bottom, one-way plow raised out of work by a cable and jib. This outfit was now sold as the Gerde D'or.

One-way or reversible plows had always been popular in Europe and several makers offered them. The Dubois motor plow was of this type with three furrows balanced on one side of the usual two

The French Latil timber tractor was employed mainly for tree felling and forestry work. This model was four-wheel drive with a ¬winch fitted at the back seen here in use.

wheels by a semi-diesel engine on the other side. The Delahaye, with its engine centrally mounted over the drive axle, used a three bottom plow at each end, one with left hand and the other with right hand bodies. This machine shuttled back and forth across the field when plowing.

Conventional style tractors included some models with narrow versions for vineyard use. The Mistral achieved this by putting the front axle on an extension in front of the radiator. This also allowed a very small turning radius. On both this and the standard model, the drive wheels were narrow with retractable lugs for road work.

SCEMIA offered a licence-built British Saunderson Universal 20. However, it was an outdated design and too expensive so was soon replaced by SCEMIA's own design. A modern enclosed style, like the Mistral, it also used a very narrow front

axle allowing tight turns; it was claimed to be able to "turn on a handkerchief". Power came from a two-cylinder vertical engine. The RIP tractor with a four-cylinder gasoline engine and friction drive was described by the English magazine *Farming* as a "ripping good tractor".

One notable French contribution to tractor design was the Latil which introduced pneumatic rubber tires to tractors. Latil had made large numbers of four-wheel drive trucks and artillery tractors during the Great War. Originally conceived in 1922 for forestry work, the tractor had both four-wheel drive and steering. Well designed and made, it was sold for agriculture, forestry and to the French army. The tires were, however, of the high pressure road variety and for off-road use, especially in mud, the wheels had lugs on the rims that folded over the tires. So successful was the Latil that the basic design lasted, under various

ownerships, until recent years.

Solid tires had been available on many industrial versions of farm tractors since around 1915 but it was not until the early 1930s that the now universal low pressure tire became available.

Unfortunately, so much of the effort in France came to nothing; the franc was devalued and fuel costs rose rapidly. The French farmer went back to his horses and oxen, and interest in tractors waned. By the late 1920s, Renault, who had revised their tractor line, showed their prototype diesel with a two-cylinder, two-stroke engine. They were now the largest French manufacturer with only the French-built Austin offering any real competition from their Liancourt factory.

Germany

The chaotic conditions in Germany following the war's end prevented farmers there enjoying the prosperity of their

French Mesmay tractor of 1920 with the popular Continental-style plow.

The Austin was Britain's challenge to Ford's model F.

counterparts in the victorious countries. Despite all this, over 30 makes were listed in 1920, although production of some was no doubt very low. The popularity of motor plows continued for a while with models from several makers including MAN and Hanomag. The latter, with eight furrows and an eighty horsepower four-cylinder gasoline engine, was the largest of them all. Rather old-fashioned with a channel iron chassis, the Pohl tractor used 25-30hp or 35-40hp engines from Kamper. Kamper engines powered other makes as well: Siemens-Sckuckert used one of their 35hp engines to power their three-wheeler with its rear-mounted, power driven rotary cultivator. Its three-wheeled layout with overhead steering made it reminiscent of the early Farmalls. Siemens had built their first model of this type in 1910 and production lasted until around 1930.

In 1924 the Fordson F went on sale in Germany and inevitably the locally manufactured tractors had to compete. Stock claimed a 30 per cent fuel saving over the F. Hanomag also made comparisons of the F with their WD Radschlepper R26. This was a far better, if more expensive, machine with a four-

cylinder, overhead valve gasoline kerosene engine of 26hp. Hanomag also made two crawlers, the Zs of 25 and 50hp, for about 10 years from 1920. The main thrust of German tractor manufacturers was towards the diesel engine and the fuel tolerant semi-diesel.

Lanz brought out the Felddank tractor which used a two-cylinder vertical hot-bulb semi-diesel. The 38hp engine was capable of running on almost anything, including melted butter! It was however, the Bulldog, introduced in 1921, for which Lanz are best known. The initial Bulldog HL, with its 12hp single-cylinder horizontal two-stroke semi-diesel engine and cast iron construction was very basic and more suited to road haulage and belt work, although they were used for field work. There was no reverse gear; the engine was stalled so that it ran backwards. Crude though it was, the Bulldog gained acceptance and was joined in 1923 by an articulated four-wheel drive model. The HL design was improved and the HR2 of 1926 set the basic pattern for future models.

Lanz would, some 35 years later, become part of John Deere, whose model D ran from 1924 until 1953. The Bulldog

The Swedish Munktell 22 built in 1922. Its specificationis included a two-cylinder, two-stroke hot bulb engine which gave 22hp.

lasted until 1960, the last diesel ones being sold as John Deere-Lanz Bulldogs. The Bulldog was, during its life, possibly the most copied design in tractor history.

Diesels

The true diesel came to the tractor in 1923 when Benz, who had been building 40 and 80hp gasoline engine Land Traktors since 1919, unveiled the Benz-Sendling S7. This 30hp two-cylinder vertical engined tractor was a three-wheeler with one rear drive wheel plus small rear outriggers for stability. A four-wheeled model, the BK, soon superseded it and was produced until 1928. Benz joined Daimler in 1926 and used the Mercedes-Benz name.

Deutz were also working on a diesel tractor and their first one, the MTZ 222, similar in appearance to the original Bulldog but with a full diesel engine, came out in 1926. It was replaced two

years later by the MTZ 220 powered by a two-cylinder cross mounted diesel. This tractor was sold across Europe.

The years leading up to World War II would see many more German tractors go into production. The main problem with diesels of this era was getting them started. The high-output battery of today had not been developed and hand cranking, aided by a decompressor, was the usual system. At the other end of the scale, the light mower tractor of the type manufactured by Deering and McCormick at the turn of the century made its production appearance, with the Fendt being one example.

Other parts of Europe

The decade following the Great War saw the growth of tractor production in many other countries in Europe. In the north, Swedish manufacturer Avance was offering two-cylinder hot-bulb semi-diesel

tractors in 18-22 and 20-30hp versions by the end of the decade. These used compressed air starting with electric glow-plug and battery as an option. This was one of the first electrically started diesels available.

The other Swedish tractor manufacturer that would later become part of the Volvo empire, Munktells, also initiated production of a two-cylinder hot-bulb semi-diesel, the 22 in 1921. This engine was derived from a small boat engine and

used compressed air from a built-in tank to start it up after an integral blow-lamp had heated the hot-bulbs. By 1930, the larger 20-30 was available and both tractors took part in the British trials of that year.

Czechoslovakia, independent since the demise of the Austro–Hungarian Empire in 1918, had two motor plow makers within its borders: Skoda, formerly Laurin and Klement who were makers of the Excelsior motor plows, and Praga. Skoda brought

The Fiat 702 which with its cast-construction, sold extensively after the war in Italy and other countries.

Manufactured at the factory in Turin, Italy, this Fiat was imported into Britain around 1919 for the World Tractor Trials. Its price in 1919 was £525.

out a Fordson style tractor in 1926 called the 30HT. The four-cylinder engine could run on either kerosene or an alcohol-gasoline mixture called dynalkol. The drive was via a conventional clutch and three forward and reverse gearbox. The following year, a smaller two-cylinder model joined it in production. The same year, Praga introduced a three model range of the AT25, KT32 and the U50. Wikov emerged as the third Czech tractor manufacturer at about the same time.

At a Czech government trial in 1927, a Skoda 30HT tractor took first prize against seven others including the inevitable Fordson F and a John Deere D among the Americans present.

The newly independent states in Eastern Europe included Poland and Hungary, where there is evidence of the International Harvester Titan being produced in the mid 1920s.

The English firm of Clayton and Shuttleworth, thresher and steam traction engine builders, had established a joint company in Budapest before the Great War with Messrs Hofherr and Schrantz. In 1923, they offered a Lanz-based model that was progressively improved and exported under several names: Le Robuste in France and the HSCS Steel Horse in Australia are two examples. It was also offered in Britain in 1930 for £350 FOB.

Limited production started in Austria with Steyr offering an 80hp tractor in 1928, although it is doubtful many were made. The Swiss also had tractors by 1929: the one-cylinder gasoline powered Hürliman IK10 and the Bührer, a four-cylinder gasoline tractor that resembled an automobile.

Italy too came on to the tractor scene with several widely different makes. The Fiat 702 with its Fordson F-style cast construction was sold quite extensively, not only in Italy but abroad as well. The improved 703 gave way to the 700 in 1926 and continued to be produced until World War II.

Bubba, Landini and the O M Testa Calda were all based to some extent on the Lanz Bulldog. The Cassani was a more advanced design using the manufacturer's own two-cylinder diesel of 40hp at a modest 450rpm. This was horizontal, very similar in appearance to a John Deere D, but with a six forward and two reverse gearbox in place of the simple high-low and reverse of the D. Production of the Cassani ran from 1927 to 1932.

For those farmers who were not diesel minded, Breda produced four-cylinder gasoline models from 1920 similar to the Fiats with 26 and 40hp engines. There were also some other minor makes in various styles.

A concept that was far in front of its time was the Pavesi-Tolotti P4 with a 20hp two-cylinder engine, four-wheel drive and articulated hinged steering. First offered in 1917 it was available until 1930. The larger P4M with a 40hp, four-cylinder engine was in production from 1917 until 1940. These lacked hydraulic assistance to the steering and rubber tires that would come when the idea was revived many decades later. They were, however, sufficiently popular to stay in production, and in the 1930s when the Pontine Marshes were reclaimed by Mussolini's government, a fleet of them were employed for plowing and other vital land work.

Around the world

European production during this time was on a modest scale. Differences in farming, language and many other things, not least monetary exchange rates, prevented much export within Europe itself. Manufacturers also had the volume-produced American imports to compete with. IHC, for example, was well represented in Europe, including Great Britain.

The revolution in Russia in 1917 followed by the Civil War stopped any progress with mechanical farming. The appalling loss of human life and the destruction caused by the Civil War left farms in no condition to feed the new Soviet Union. The victory of the Communists in 1920 allowed a limited start to tractor home production. Several prototypes were built and tested. The time to perfect them was simply not available and imports and licensed production looked to be the best way forward. After testing various makes the first Kommunar crawler tractors, based on the German Hanomags, came off the production line from a factory in Kharkov along with Fordson Fs from the Putilowitz factory.

From the early days, Soviet tractor production had close ties with the armed forces who used considerable numbers of

An early 1920s Fordson with an in-line four-cylinder engine.

them, especially crawlers. Soviet technology and production quality standards were a major problem for many years and reliability was often very low. The new Soviet government under Joseph Stalin set out to forcibly collectivize farms into huge state-run enterprises that would require tractors and associated machinery on a vast scale. One of the beneficiaries would be the American tractor industry which would receive its biggest orders ever.

On the wider world stage, Canada lost its home production. Sawyer-Massey were Canada's largest steam traction engine and thresher builders and like many in the industry they had gone into the tractor business with large tractors to replace

their steam traction engines. From 1910 until the early 1920s, Sawyer-Massey built a range from 27-50hp down to an 11-22 introduced in 1918. The Brentford, Ontario company of Goold, Shapley and Muir also left the traction industry after about 12 years. When Canadian tractor production resumed over 20 years later the new manufacturers would introduce a revolutionary new feature.

Australia on the other side of the world lost one manufacturer and gained two. Jelbart ceased production and Ronaldson and Tippet offered a tractor generally considered to be based on the American Illinois Superdrive that had recently ceased production. One of the Bulldog's pups reached Australia in 1929 when

The Crawley Motor Plough from England dated 1920. The prototype of this machine was developed in 1908 by a farming family. A factory was eventually established in the East of England in 1914, and production lasted until 1924.

McDonald's introduced their Super
Diesel with a Lanz-style engine fitted
on to an imported Rumely chassis and
transmission; later when Rumely ceased
production, the whole tractor was home
produced. There had been several other
attempts at an Australian tractor but they
had all fizzled out.

About the only thing now missing from
tractors was the three-point draught
control and hydraulic lift, and even that
was in prototype form. Harry Ferguson
had followed up his earlier work on the
Eros car conversion by transferring the
mounted plow on to the Fordson F. When
this had reached a reliable working stage,
Harry Ferguson tried to interest various
companies to manufacture it for him to
sell. The job fell to an American company,
Roderick Lean of Mannfield, Ohio.
Unfortunately, they went bankrupt in
1924. Ferguson then joined forces with
the Sherman Brothers, Fordson dealers
from Evansville, Indiana.

Harry continued with his experiments
back in Britain and eventually decided on
hydraulics as offering the best potential
for what he had in mind. The first
prototype was built on to a Fordson F.
Unfortunately for Harry the F was by now
out of production.

History was about to undergo one of
its periodic events that would affect the
future of everything, including the tractor.
In October 1929, the United States Stock
Market crashed, with devastating results
on the economy and industry.

The Ferguson Duplex plow circa 1922
mounted on a Fordson F, possibly in
Lincolnshire, England. Harry Ferguson can
be seen on the far left.

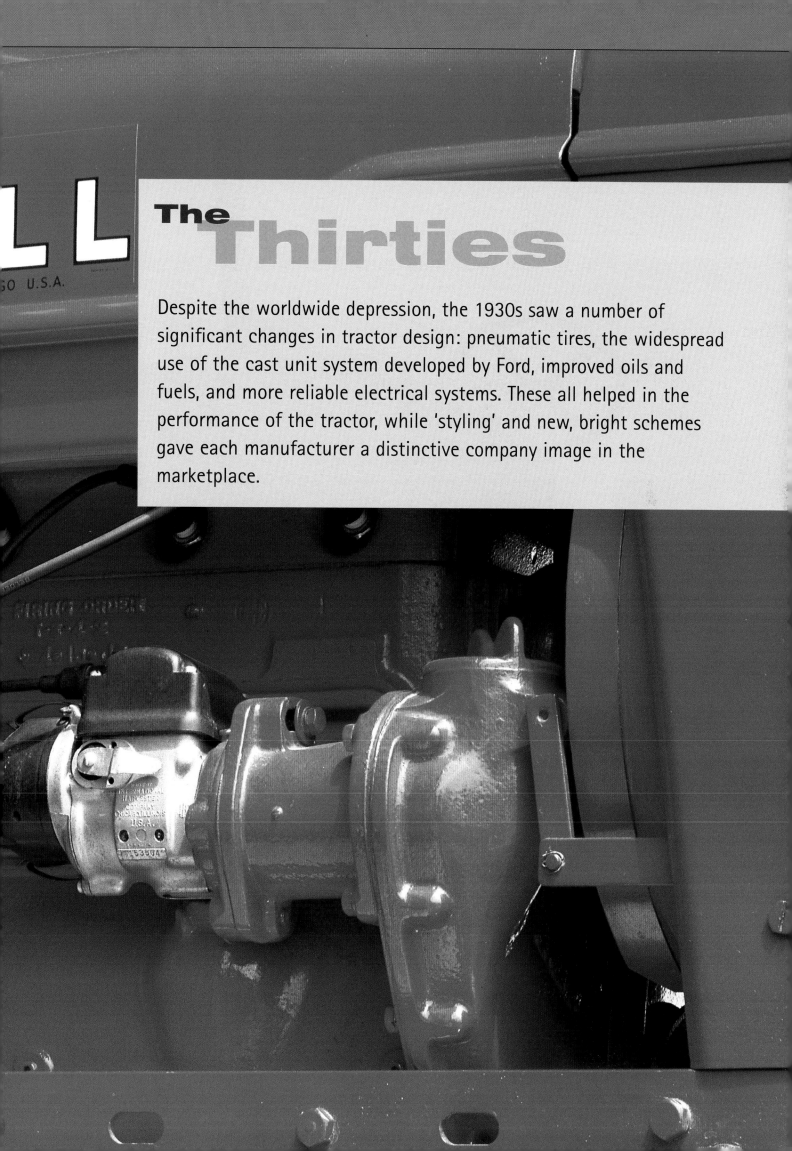

The Thirties

Despite the worldwide depression, the 1930s saw a number of significant changes in tractor design: pneumatic tires, the widespread use of the cast unit system developed by Ford, improved oils and fuels, and more reliable electrical systems. These all helped in the performance of the tractor, while 'styling' and new, bright schemes gave each manufacturer a distinctive company image in the marketplace.

The 1936 Oliver 28-44, still bearing the Hart-Parr name underneath, a practice which ended shortly after this model came off the production line.

The Thirties

The start of the new decade in America was in sharp contrast to the heady days of 1920 with its multitude of makes and models. Most of the sales were now divided among nine major manufacturers: Allis-Chalmers, Minneapolis Moline, Massey Harris, Case, Caterpillar, Cletrac, International Harvester, Oliver and John Deere. Of the smaller ones, only a few survived until World War II: Baker, Eagle, Huber and Keck–Gonnerman, with Rock Island being taken over by Case in 1937 and Bates finishing the same year.

The new models from all manufacturers reflected the experience and engineering progress at the time. Ford's cast unit system became universal except on some row-crop tractors where the need to attach mid–mounted equipment was made easier by a channel frame. Improved hardening and heat treatment of gears, crankshafts and axles prolonged their working lives. Oil bath air cleaners, oil filters with replaceable elements, along with better oils and pressure oil systems played their part in keeping tractors at

Massey Harris four-wheel drive tractor manufactured between 1930 and 1936. It housed a Hercules four-cylinder side valve engine developing 25hp.

Pneumatic tires

The thirties' greatest contribution was the low pressure pneumatic rubber tire which turned the tractor into the truly universal farm horse replacement it set out to be. Gone were the awful jobs of fitting roadbands to the wheels to travel on tarmac roads and removing spade lugs to prevent pushing hay into the ground at hay–time. The farmer could now use the public roads to reach his distant fields and haul his produce to town.

Solid rubber tires had been used on tractors for road haulage and industrial work, but they had failed to work on the land. Latil in France used clip over lugs on high pressure tires in bad conditions, but this had only been a stop gap solution.

Florida fruit growers sparked off the tractor tire as we know it today. In the late 1920s they had become concerned about root damage caused by spade lugs. Some owners had bolted large, used truck tires to their tractors while others tried aircraft tires.

The results were sufficiently successful for some tire companies to take notice. B F Goodrich tried a zero pressure tire but it was Firestone who came up with the answer; modified aircraft tires with angled lugs molded on and inflated to 15psi, a figure that has remained the average to this day.

Of the tractor manufacturers, it was Allis-Chalmers who saw the tire as the item needed to boost sales. The U - as it was now called - was

offered with tires in 1932. Publicity for the new tires consisted of long road runs, a well-known tractor sales stunt, and a new one: speed runs. Allis employed some well-known racing car drivers to drive Us in races at agricultural shows and through towns in the hope of being stopped for speeding by the police. One driver, Ab Jenkins, used a U with a modified drive line to raise the world's tractor speed record to over 66mph. These stunts finally convinced farmers that tires were a workable and worthwhile part of tractors.

A 1938 Case RC model on display for potential buyers. The sign at the front of the tractor lists the machine's specifications and options.

The successful Allis-Chalmers B model introduced in the late 1930s in the new livery of Persian orange.

A 1936 Hart-Parr Oliver 70 row
crop tractor.

work. Electrical systems became more
reliable and electric starters were offered
as options along with lights that enabled
farmers to catch up in a late season by
working at night, something you could
not really do with horses. Refinery

techniques were now able to provide fuel
of a constant and better standard of
consistency. Kerosene, the popular early
fuel, was still available but distillate, with
a higher rating came on the market, as
did 70 octane gasoline. British equivalents

were paraffin for kerosene and vapourising oil for distillate. The European countries used a variety of descriptions for different fuels.

One thing that did not improve the tractor's performance but did improve its looks was styling, introduced in the late 1930s and used to give a product an individual image that people remembered.

The Allis-Chalmers line at this time consisted of the 40hp E, replaced in 1936 by the A, the U, and its row-crop version, the UC, plus several crawlers. The U models received Allis-Chalmers' own UM engine in 1936. The WC model joined them in 1934 as the first tractor to be designed for rubber tires with steel wheels as an option.

The crawlers were descended from Allis' takeover of Monarch and provided farmers as well as the increasing industrial and earthmoving customers with tractors up to the 75dbhp of the six-cylinder model L. Allis-Chalmers made its first foray into diesels with the Waukesha-Hesselman engines in some LO and SO models. The Hesselman was a compromise engine with the compression of a gasoline engine, diesel fuel injection and spark ignition. At the other extreme, Allis in 1938 offered the small B on tires only. This was a real winner for Allis-Chalmers, and after World War II it was also made in Britain both for sale on the domestic market and for export. 1938 also saw the styling of the WC and its standard version, the WF to match the B. With their Persian orange color scheme introduced with the U, they were a smart line up. Curiously, the U was not styled, like some other makes, for example the Oliver 80; the use of a large cast radiator rather than an automobile type may have made it too expensive an exercise to undertake.

Cletrac, with one exception, made only crawlers, from the small row-crop 15 with 17dbhp to the massive 80G capable of around 100hp. They foresaw the diesel's potential and in 1933 put a six-cylinder Hercules diesel of about 90hp into an 80 tractor. The number series gave way to the use of letters from A to F on both spark ignition and diesel engined models. The out of sequence E was the smallest at under 20hp, the largest being the FD at 100hp. In 1939 they were joined by the even smaller HG crawler and Cletrac's only wheeled tractor, the GG General, basically an HG in three-wheeled row-crop form. In the early 1940s, Cletrac passed this tractor to B F Avery of Louiseville, Kentucky. From the first,

1939 was the final year of the Case gray. This model C 17/27hp tractor was later superseded by the D series styled in Flambeau red.

Cletrac used electric start on their diesels and indeed from about 1930 offered them on all their crawlers. This, combined with the controlled differential steering using band brakes that allowed both tracks to be driven whilst turning, helped Cletracs to become a popular tractor among farmers and industrial users. The tractors were styled around 1937 with the radiator grill that was to stay with Cletrac until the end of crawlers under Oliver ownership almost 30 years later.

Once the dust of the earlier merger had settled, Caterpillar started to fill out its range of tractors, and by 1930 had the Ten, Fifteen and Twenty tractors in addition to the Thirty and Sixty from the merger. Caterpillar had been working on a diesel engine for a while and in 1931 a diesel engined Sixty was announced. Nebraska tested the tractor the following year and it gave 54dbhp at a modest 700rpm. Once the diesel had been accepted, Caterpillar produced diesel versions of most of the range. The range was, like Cletrac's, capable of covering almost any farm or industrial requirements.

By the mid 1930s, Caterpillars were beginning to assume the look that would make them recognized the world over in the coming years. The big diesel 75 of 1933 was the forerunner of the most famous crawler of all time, the D-8, possibly only equalled by its smallest relative, the D-2 that came out in 1938. The spark ignition engine faded out, except as the pony starting engine fitted to all Caterpillar diesels until comparatively recently.

The cross motor style went out in 1929 when Case went conventional with the introduction of the L standard tread and the C in standard and row-crop versions.

These became very popular and long-lasting tractors. An innovation on the CC was the motor lift for the tool bar; this saved the driver a lot of effort at the row ends. A smaller R and RC came out in 1936. In keeping with the industry trend, Case changed its color scheme in 1939 from gray to Flambeau Red; at the same time the C series was replaced by the styled D, and the R was styled to match, leaving only the L untouched until replaced by the styled LA in 1941.

The D mainstay of John Deere was continually uprated and improved so that by the late 1930s it had more power, three forward speeds and presented the farmer with an excellent tractor. Row-crop needs were met by the A in 1934 with, for the first time, a hydraulic lift for the tool bars. This was also available on the smaller B the following year. Deere styled the A and B in 1938 and introduced their largest and smallest row-crops, the G and H in the same period. Unusually for Deere, their other new model, the very small L, had a vertical twin engine instead of the horizontal twin they used in all their other tractors.

International Harvester

The faithful 10-20, 15-30 and Farmall Regular that had brought IHC through the 20s were improved as the new decade got underway. The Farmall Regular became the F-20 and had a larger F-30 and smaller F-12 offered alongside it to cater for all sizes of farm. The 15-30 was uprated to 22-36 in 1929 and was replaced in 1934 by the W-30. However, the 10-20 remained in production until 1939, although according to Alan T Condie in his excellent series on British tractors, it was coming into Britain until 1941 under the Lend-Lease agreement.

The International Farmall F-12 built in 1933 for smaller size farms.

1937 witnessed the arrival of the Minneapolis Moline model Z tractors in the bright new color - Prairie Gold. Shown here is the ZTS.

IHC changed from gray to red for its tractor livery in the mid-thirties. It has been suggested that with the coming of the rubber tire, more tractors were using the public roads and companies like IH, Case and Minneapolis Moline changed the color from gray to make their products more visible to other road users.

The first American wheeled diesel tractor came from IHC in 1934 in the shape of the WD-40. Its four-cylinder, 44 belt horsepower engine started on gasoline for a warm-up period before changing on to high compression and diesel fuel by the closing of a third valve in the cylinder head. This rather complicated system, which enabled the engine to be hand started, was used by IH for about 20 years. A six-cylinder spark ignition engine was the alternative in the W-40 model.

To farmers and industrial users who wanted crawlers, IHC offered their TracTractors. The smallest was the T-20 with spark ignition engine only, and the larger T-35 and T-40 which had the split head diesel as an option. Unusually, the steering clutches on these tractors were in line with the gearbox rather than the track sprocket axles, and could be serviced from the rear of the tractor. The entire range was cleared out in 1939-40 and replaced by a new one with a universal style that encompassed a small row-crop at one end and a large diesel crawler at the other. They were to become one of International Harvester's most popular tractor ranges and the style could be seen in modified form right into the 1960s, not only on American-built tractors but IH tractors worldwide.

Pacemaker and Challenger were the names Massey-Harris gave to their revamped Wallis tractors, of which the Challenger was the first row-crop tractor to use the boilerplate frame. They were given another facelift and an improved engine mid-decade together with a new color scheme, a red body with straw yellow wheels. Top of the line was the Model 35 which got the same treatment and lasted until 1940. In contrast to their rather dated models, Massey-Harris brought out a four-wheel drive general purpose tractor in 1930 with kingpin steering and a variety of wheel track widths to suit different row-crops. It did not sell as well as expected and was re-engined with a valve in head engine in an effort to make it more popular. However, even with rubber tires it had faded out by 1939. Another innovation from Massey-Harris was Twin Power. This was an override on the throttle that gave extra engine speed for the belt pulley. The first of a new line was introduced in 1938: the 101 Super tractor set the tone for Massey-Harris right to the end of their final models.

A new name on tractors was Minneapolis Moline, based on the Twin City tractors from the amalgamation of 1929. First with the new name was the Kombination Tractor in 1930; a four-wheel row-crop with a four-cylinder 11-20hp engine. The Kombination Tractor got its name because it could handle a combination of field and row-crop work as well as pulley and power take-off jobs.

For tricycle tractor buyers, the MT Universal was offered with an engine that could be set up to use gasoline, kerosene or the new distillate fuel. The standard tractor in the line was the FT21-32 from the old Twin City line. In 1935 they were all updated as the KTA, MTA, and FTA. These, along with the Universal JT of 1934 in row-crop or standard form, started

Minneapolis Moline's reputation of long life and reliability.

A new color, Prairie Gold with red wheels, came with the model Z tractors in 1937, followed by the U the next year along with the big new GT Standard. The Z engine was very unusual in that the valves were horizontal and the compression ratio could be altered via a metal insert that protruded into the cylinder head opposite the valves. This enabled the engine to be run on different fuels. The Deluxe UDLX with cab and front and rear fenders was Minneapolis Moline's last pre-war model. This type of fully enclosed, bodied tractor had been popular in Europe, particularly Germany, for many years but history has not recorded where Minneapolis Moline had studied the concept before trying it out on American potential buyers.

Alphabetically last, but by no means least, the new Oliver tractors started a line that would be noted for innovative and hard working machines. Prior to the merger, Oliver Chilled Plow had built a series of prototype standard and row-crop tractors and the new versions looked a lot like them, although Allan T Condie, in his Oliver book, suggests that some of the design team may have formerly been International Harvester people and he

A 1937 model of the Hart-Parr Oliver 70 row crop tractor. This machine (left and far left) was carefully restored by its owner Robert H Tallman of Delaware, United States.

notes the similarities of the new Olivers to IH tractors. Initially, to give them continuity to the company's predecessors, the Hart-Parr name appeared with Oliver on the new 18-28, 28-44 and Rowcrop tractors. The Rowcrop used the very narrow rimmed rear wheel with offset lugs developed by the experimental Oliver Chilled Plow tractors: the tiptoe wheel became a well-known Oliver feature. As with other manufacturers, Oliver updated their tractors in the middle of the decade and the models became the 80, 80 Rowcrop and 90.

To take advantage of the new 70 octane gasoline, the Oliver 70 in standard and row-crop models came out in 1935, a low compression version, the 70KD (kerosene/distillate), followed later. The 70 was Oliver's first six-cylinder tractor and used a fully enclosed engine with louvred side panels; a very smart outfit. For some reason Oliver never styled the 80 and 90 series and these tractors remained the same up to the end of production.

Depression

Dust bowl, depression and a very competitive marketplace did not entirely sweep away the small companies or deter new ones from entering the industry.

Eagle from Appleton, Wisconsin, built their 6 series in small numbers using six-cylinder engines from Waukesha and Hercules, and Huber's Super Four models continued in production. In 1937, along with the Modern Farmer row-crops, Huber produced a very stylish tricycle model, the B with a Buda engine of 27 belt hp.

One novel tractor of the depression era was the economy model using secondhand car parts. Thieman Harvester Co was one of a number who supplied kits to farmers. Car conversions had first been used during the First World War to meet the huge demand, now in a time of lowered demand they were used to keep costs down.

First of the newcomers was the Plymouth from the Fate-Root-Heath Company of Plymouth, Ohio. It seems the

Harry Ferguson (left) together with Henry Ford senior and the fruits of their agreement - the all new 9N tractor. The historic deal between Ferguson and Ford in 1938 was sealed with simply a handshake.

Plymouth car manufacturers requested a change of name, so in 1936 the name Silver King came into being. The tractor was a small standard or tricycle. It was locally popular and continued well into the 1950s.

An assembled tractor of larger size and noted for its high top speed was the Co-op, available in three models. Built by a variety of companies over its production life, the Co-op tractors lasted for 20 years.

The GB 503 tractor was assembled by the Graham-Paige Motor Corporation and offered for sale by mail order company Sears Roebuck.

Ford's return

Henry Ford, scourge of tractor companies in the post-Great War years, returned to the scene in 1939. Harry Ferguson, whose tractors had previously been made by David Brown in Britain, forged a new deal with Henry Ford senior in the US; the two men simply shook hands and the gentlemen's agreement was sealed. The eventual outcome of this deal was the all new 9N tractor. This owed little to any previous tractor except the Brown-built Ferguson A of 1936-1939 (see Hydraulic Lift below). In reality it used a lot of available Ford vehicle parts, modified

where necessary, to keep costs down. The four-cylinder engine was essentially half a Ford V-8 producing 20hp on gasoline, for export mainly to Britain. Under the Lend-Lease arrangement the 9NAN low compression version was produced. The 9N model was a utility tractor; a standard style but with adjustable track and row-crop capability.

At the decade's end there were - to all intents and purposes - ten American tractor manufacturers, a number greatly reduced from 47 in 1929 and 186 in 1921. All had new models either in or ready for production; this was fortunate because the world was once more about to be engulfed in war.

British trials

What happened during the decade in other parts of the world? In Britain, 1930 witnessed the inauguration of the grandly titled World Tractor Trials at Benson in

Oxfordshire. United States entrants were the Case C and L gasoline kerosene; the Caterpillar 10, 15, 20, 30, and 60 all gasoline; the IH 12-20, 15-30 and Farmall all gasoline kerosene; and the Massey Harris 12-20 and 20-30, also gasoline kerosene. All performed well on both belt and drawbar jobs, although the Caterpillar's high fuel costs told against sales. Caterpillar did, however, later offer kerosene engines on the farm crawlers.

The British entries were a very mixed bag. The Vickers' Aussi was withdrawn during the trials because of the decision to close production, and the Peterbro broke down. This tractor was also about to go out of production. The Fordson N, listed as an Irish entry, suffered a cracked block and withdrew. Another make with a shaky future was the Rushton, one on wheels and another on Roadless rubber jointed tracks. Originally called the General, the Rushton was first offered in

Ferguson's 1932 'Black' tractor shown here with a three row ridger. First produced from the Belfast works in 1933, the tractor weighed only 16.4 cwt, housed an 18hp Hercules engine, and a David Brown gearbox. The implements could be attached or detached in under a minute and the driver could set the controls with ease from his seat.

The Case L of 1929, on trial at the first British World Tractor Trials held in Oxfordshire in 1930.
Case was one of several United States entries.

late 1928 by AEC Limited who were behind the project. It was so closely based on the Fordson, that parts would fit either make and major components could also be interchanged. It is said that Ford had a quiet word about this closeness and after some redesign the Rushton emerged. The depression, together with competition from the Dagenham Fordson N, and an unpaid order from France finally finished Rushton in 1937.

The four other British tractors were all diesels. McLaren's of Leeds, builders of a variety of steam machinery, entered their licence built Benz two-cylinder diesel, practically all of which went for export. The same company also entered a genuine Benz with a single-cylinder horizontal engine. Marshall, who had entered tractors in the Winnipeg Trials before the Great War, had long been absent from the tractor scene. They had looked at the Lanz Bulldog and other Continental tractors and had decided that the full diesel was the engine to choose. Their first one, the 15-30 with its 8 inch bore and 10 inch

Marshall's 15-30 model of 1930 had a single-cylinder, two-stroke diesel engine. It proved to be the forerunner of the Field Marshall, produced after World War II.

stroke horizontal, single-cylinder two-stroke full diesel was the ancestor of the Field Marshall of post-World War II. The Marshall again was hand started with the aid of a special igniter paper that was lit and then screwed into the cylinder head in a holder. The modified 18-30 model trickled out during the mid-thirties. The 12-20, a smaller tractor, later called the M, came out in 1934 and ran until 1945.

Two very advanced diesel tractors came from A.G.E who were on the verge of collapse. Agriculture and General Engineer had been created in 1919 to combine a whole range of companies, many of them being steam traction engine and associated equipment builders, into a single, hopefully world beating company. It was a flop, just like the American United conglomerate. Garretts of Leiston, part of A.G.E, built a big tractor in 1930 with a choice of diesel engines. The Blackstone engine running at 1000rpm with a 14:1 compression ratio gave a maximum of 37 belt and 27 drawbar horsepower. Blackstone, also part of A.G.E, were longtime stationary diesel engine makers and used their patent spring injection system which was not really suited to vehicle use. The better engine from Aveling and Porter of similar size and speed produced a maximum of 42 belt horsepower and 30hp at the drawbar. This one used electric motor starting unlike the gasoline pony on the Blackstone. Regrettably not many were made due to the A.G.E's collapse. They were, however, the world's first four-cylinder high speed diesel tractors and the judges commented very favorably on the performance and economy of the diesel.

Roadless Traction, who had supplied the tracks for the Rushton, had perfected a method of joining track links by

compressed rubber hinges instead of the usual pins. They converted all sorts of makes into crawlers on request and supplied two specialist crawler makers who made their debuts in the mid thirties. Bristol made a narrow crawler employed mainly for row-crop work which used an

opposed two-cylinder gasoline/kerosene engine, and Ransome's of Ipswich, Britain's largest plow manufacturer, brought out an even smaller one, the MG-2 with a single-cylinder motorcycle type engine for truck farm and greenhouse use.

Hydraulic lift

Following the end of Fordson F production, Harry Ferguson decided to build a tractor to suit his draught-controlled lift. He bought an American Hercules engine and then contacted David Brown Gear Cutters of Huddersfield, Yorkshire, for the transmission. After the prototype was tested, Brown agreed to build the tractor for Ferguson to sell, and in 1936 the world's first production tractor with modern style hydraulic lift rolled off the production line. It was called the Ferguson Model A. After

Harry Ferguson's first production tractor with modern-style hydraulic lift - the model A - left the factory in 1936. Here, a model A is seen at work with a binder. This model was built by David Brown of Huddersfield, Yorkshire.

initially using a Coventry Climax engine, later versions were fitted with a David Brown-built unit until production ceased in 1939; cost and the need for its own implements did not help sales to depressed farmers. David Brown wanted to increase the power and Harry Ferguson wanted to reduce the cost, and so they parted company, with Harry Ferguson going to seek assistance in America where he thought Henry Ford might be more agreeable to his ideas.

Meanwhile, David Brown put his ideas into practice and at the 1939 Royal Show exhibited his own tractor, the VAK-1, complete with hydraulic lift but no draught control, as Ferguson's patents were very tight. The implements used adjustable depth wheels instead.

Dagenham on the River Thames near London, England was now the site for Ford's tractor factory and, since its opening in 1932, Fordsons had been exported around the world, even back to the United States for which Ford produced their very first row-crop tractor, the All Around model in 1936. This was the first time in tractor history that American manufacturers had faced overseas competition on their home territory.

The other Fordson plant was in the Soviet Union where production ended in 1932 and the Soviet copy of the Farmall, the Universal took its place. The Soviets also put the IHC 15-30 into production as the SKhTZ 15-30 at Kharkov and Stalingrad. The S-60 gasoline and later the S-65 diesel based on the Caterpillar 60 were produced at a vast new factory at Cheliabinsk in the Urals. The Soviets produced huge numbers of crawlers for agriculture, industry and the armed forces. The Soviet Union also imported a lot of American farm tractors in the 1930s.

Diesels in Germany

The diesel engine made a real impact on German tractors after 1930 and many well known makes came into being at this time. Deutz's early two-cylinder cross motor diesels were replaced by the very modern looking Stahlschlepper or Iron Tractor models. The F1M 414 had a single-cylinder 11hp diesel. The F2M 317 with a 30hp two-cylinder in-line diesel running at 1300rpm and the F3M 315 with its 50hp three–cylinder engine would not have looked out of place 30 years later. The two smaller ones offered electric starting while the 50hp model used one cylinder to compress air ready for the next start up before running on all three for actual work.

Fahr, who would eventually become joined with Deutz, also started diesel tractor production using a 22hp Deutz engine. This was a popular size on small German farms and about ten other tractor companies were offering Deutz diesel engined tractors in the up to 25hp class. Three or four forward gears were usual although the Fahr had five forward gears and the Stock offered six plus two reverse as standard.

A company that enjoyed a thriving export trade was Hanomag, with its wheeled diesels and a crawler. The R38 used the D52 engine that was introduced in 1930 and continued in production into the 1960s for all sorts of applications besides tractors. The R50 and K50 crawler used a larger version of the same engine rated at 50hp. These tractors also had the 540rpm power take-off shaft that was becoming universally accepted. During this period a version of the agricultural tractor known as the Strassenzug Maschine - literally "street towing machine" - became widely used in

Germany and to a lesser extent in other European countries.

Most famous of these were the Lanz EIL Bulldogs of which the 55hp D2538 was the largest with a top speed of 22mph in fifth gear. With the enclosed bodywork, front and rear fenders and fully equipped cabs, they were very popular for road

haulage not only of farm produce but for timber, coal and many other local delivery services. More extreme versions of the road tractor such as the Hanomag RL20 and the SS20 looked more like cars on oversized tires than tractors. Diesel powered, they were capable of 16mph in fourth gear, their agricultural origins given away by the fact that you could still get a steel wheeled, three geared model if you wanted. At the other end of the scale, the Hanomag Strassenzug Maschine Type 100 Gigant with its 100hp six–cylinder diesel was a truck tractor unit for drawbar trailers with no land use capabilities whatsoever. Most common makers offered

The popular Lanz Bulldog. This model dates to 1937.

a street version with varying degrees, enclosed bodywork and a high top gear of around 20mph.

Lanz had by now a range of Bulldogs including the 55hp model T crawler, the wheeled models being the 15hp L, 23hp N and 45hp P. They were sold in Britain by Lanz Tractor Co Ltd of London who described them as crude oil engined at prices from £295 to £450 on rubber tires and £825 for the crawler. The term crude oil emphasized the fact that they would burn just about anything. One sales catalog suggests that the farmer makes friends with his local bus or haulage company to obtain their used engine and gear oil. After filtering, this was mixed with 25 per cent kerosene and used as fuel, costing nearly nothing – this was a big plus point in the depressed 1930s. Another unusual feature was the starting method: after heating the hot bulb with a blow lamp, the steering wheel and the top part of the column was removed and used to swing the engine over; the lever steered crawler even carried a steering wheel just for starting up!

Renault's 1933 model VY diesel produced until 1939.

RENAULT Y.L 1933

Simple tractors

In contrast to all the advanced tractors in production there was also a market for very simple tractors, some no more than stationary engines with a very simple transmission and chassis. Examples of this are the Fendt Dieselross (diesel horse) and the Herman Lanz Samson (no relation to the other Lanz). A tractor that fell in the middle was the Mercedes Benz OE with a single-cylinder, horizontal, hopper-cooled four-stroke diesel engine of 20hp, an example of which had been at the British Tractor Trials of 1930.

However, not all German tractors used diesel engines, and both gasoline and gasoline/kerosene engines were made in small numbers. Around 25 tractor companies were building tractors, but by 1939 production was being measured in hundreds rather than thousands.

French tractors

Tractor production in France was virtually in the hands of Renault and Austin. The PE series of gasoline and gasoline/kerosene tractors (still with the now-vertical radiator behind the engine that had been introduced in 1926) continued until 1937. Renault had been experimenting with several diesel models and in 1933 brought out the VY with a 30hp four-cylinder diesel running at 1200rpm. The radiator was now at the front and the engine was fully enclosed with its smart yellow and gray color scheme; this was France's first volume produced diesel tractor. The gasoline YL of 20hp and 55hp VI diesel crawler completed the line of tractors. The old P series was replaced by the AFV gasoline in 1938 when the AFX was added together with the big 85hp AFM crawler. The AFV with a 540rpm p.t.o shaft and independent brakes was also available in vineyard form. The vineyard tractor was to the French farmer what the Farmall was to their American cousins, and both wheeled and crawler tractors were available from French manufacturers.

Austin's original French tractor was initially the same as the British model. However, the tractor entered in the 1930 British trials was much improved with a three speed gearbox and many other improvements including a p.t.o. The kerosene model gave 19 belt horsepower at maximum engine speed and the straight gasoline gave 24hp. Narrow vineyard versions were also offered and from 1937 a diesel 16-28. Production was terminated early in World War II.

Citroën-Kegresse and Latil fulfilled special needs along with a small number of vineyard tractor makers. One of these, Educa, prototyped a crawler machine with rubber tracks running on rubber tires, very similar in concept to the Caterpillar Challenger of 50 years later.

The Société Français Vierzon brought out their version of the Lanz Bulldog in 1935 with their models the H1 of 22–38hp and the H2 15-25hp. With styled bodywork in green with yellow wheels and enormous exhaust and air cleaner pipes they were formidable machines. Uprated in 1937 to 38-44 and 25-30, they were joined by the even larger H0 45-50 in 1938. On rubber tires, the buyer had a choice of three or six speed gearboxes. S.F.V would be taken over by J I Case in about 1960.

Across Europe

The Alps and other mountain ranges of Europe meant that there were very many small farms which were hardly conducive to tractor farming methods let alone manufacture especially in the economic climate of the 1930s. Swiss makers Hürliman brought out the 2M20 in 1934 with a 20hp two-cylinder engine and then in 1939 their high-speed direct injection four-cylinder diesel. AEBI, who had made

The Swiss Hürliman Diesel tractors were generally designed with small wheels and a low center of gravity for safety on steep sloping fields.

The Swedish BM 2 tractor manufactured in 1939 with a two-cylinder engine producing 31hp.

self-propelled mowers for mountain farmers for a number of years including a four-wheel drive model, added a small tractor to their range. Swiss tractors were designed with small wheels and a low center of gravity for safety on steep sloping fields.

A company more associated with armament, Bofors of Sweden, joined Avance and Munktells in offering a two-cylinder semi-diesel, the 40-46, in 1932. A crawler version was also offered and a number were used by the Swedish army. Munktells joined forces with their engine supplier in 1932 to form Bollinder–Munktell, and continued their 15-22 and 20-30 semi-diesels.

The largest Italian maker, Fiat, added

their first crawler to their line in 1932, the 700C, based on the wheeled 700. This wheel-steered 30hp four-cylinder kerosene model was joined by the 708C of 20hp in 1934. The model 40 Boghetto introduced in 1939 was in production until 1947.

In the same way that General Motors had tried to take on Ford with a tractor, Fiat's rival Alfa Romeo brought out a superior tractor to Fiat's 700. Powered by an overhead camshaft engine it did not last much longer than the Samson did against the Fordson F.

Breda made a conventional style four-cylinder 40hp gasoline kerosene tractor and a diesel using a two-cylinder, two-stroke Junkers engine, a very unusual type

Lanz copies

Italian tractor makers went in for Lanz copies in a big way. The Bublio UT4 of 1930 and the UT6 which replaced it in 1937, along with the C and later the CA crawler, bore the Lanz style except for the conventional radiator mounting at the front of the engine. This was also a feature of the Landinis built under Lanz licence as the 30hp Vélite, 40hp and 50hp SL50 Super L. The 30 and 50hp were made right up into the 1950s and were very popular in Italy. Another Lanz licence went to Orsi who offered almost identical tractors as the Artiglio, Super 43 and Super 50. There was at least one other Lanz based tractor, the Deganello Sabaudia produced from 1929 to 1938.

Czechoslovakia had been building its own power plows before World War I. These were the forerunners of the Czech tractor industry. In 1926, Skoda came up with the 30 HT farm tractor which became a huge success in eastern Europe. The model shown here - the Wikov 22 - was built in 1937 by the Wichterie and Kovarik company of Prostejov, a well-known farm machinery manufacturer. It houses a gasoline/kerosene two-cylinder engine and could offer steel or pneumatic tires.

The four-cylinder Cockshutt 80 was the
Oliver 80, rebadged for sale in Canada.

of diesel. OM was another producer offering a 48hp gasoline/kerosene tractor, the 2TM from 1938. They had also built a Lanz-based model ten years previously. Motomeccanica's Balilla 10hp wheeled and 15hp crawler tractors with four-cylinder engines entered production in 1930 and lasted until 1945. Their S50 crawler had a Hesselman four-cylinder engine of 50hp. Offered from 1935 to 1945, it had a six forward, single reverse gearbox whereas the small Balilla wheeled tractor was offered with six forward gears plus two reverse, similar to the Balilla wheeled tractors.

Cassani's diesel succumbed to the recession in 1932 and the company went into motorized mower tractors with 610cc gasoline engines under the name Same in 1936.

The final European model of the decade was a newcomer from Hungary, the Mavag, a conventional tractor similar to the Case L.

Tractor production in Australia consisted of the McDonald Imperial Super Diesel in low numbers and the Ronaldson-Tippet 20-36 until the depression finished it in 1937. G R Quicke, author of *Australian Tractors*, gives about 340 as the total production figure. This is not many, but Australia has always required specialized tractors matched to its unique needs and there have long been many low-volume producers catering for this market over the years.

Harry Ferguson wasn't the only one affected by Ford's decision to stop F production in 1928. Cliff Howard, an Australian who pioneered the Rotavator rotary cultivator had been making Rotavators to suit the F. He also made the p.t.o drive and special gearbox for the transmission to get the forward speed he

wanted. Just as Harry and his plow were left high and dry, so was Cliff Howard and his Rotavator. His reaction was the same: he built a suitable tractor himself. The DH22 with its distinctive vertical steering column was powered by a 22hp gasoline/kerosene engine driving through a ten forward speed gearbox, probably an industry first. Later models had rubber tires and hydraulic lift, and the DH22

enjoyed a thirty year production run.

Regrettably, the tractor's fiftieth birthday was celebrated in the same manner as its 25th, by going to war. In the Second World War, the tractor would once again play a major role both in the fields and with the fighting forces. On the 1st September 1939, Hitler's forces stormed into Poland and within days Britain and Germany were at war.

The quest for fuel

Fuel was a constant worry in Europe because practically all fuel had to be imported. It was often subject to high sales taxes; gasoline in Britain being the classic example. A lot of work was done on alternative fuels: alcohol mixtures were one avenue of research, the other possibility that occupied a lot of effort was the producer gas system. A slow burning wood fire in a firebox carried on the tractor produced gas that was passed through a cleaner before being used in the engine either neat or mixed with liquid fuel. The French were very much at the forefront with this technology in the 1930s, which was fortunate because they found that this was often their only source of vehicle fuel during World War II. The depression in Australia gave rise to the same quest for a cheap home produced fuel and a lot of farmers built their own units to fit to their tractors for producer gas operation. G R Quicke in Australian Tractor gives a detailed description of how the gas producer works in both theory and practice.

Above: A wood-burning tractor from Munktell's 1938 line. The two-cylinder engine was converted to produce gas from the burning wood.

Eimco Power Horse

This beautifully restored machine was originally produced in 1937 by The Bonham brothers - Bond and Bert - together with the Eimco Machinery Corporation of Salt Lake City, Utah.

The Power Horse filled a need on the small farms and ranches in the western United States as it literally replaced the horse. It was designed to be controlled completely from the seat of a mower, dump rake, wagon or sulky plow. Just snap the reins - like with a horse - and this amazing model closes the two control clutches and the machine takes off; pull on the reins, and it stops. Pull extra hard and the tractor will back up; release the reins and it goes back to neutral. Pull one rein back to reverse and the other in forward and the tractor will spin a turn in its own length.

This tractor is four-wheel drive and weighs in at 2500lbs. Roller chains drive the wheels in the large side castings. To this tractor's gear box, transmission and wheels, an Allis Chalmers B engine was added, giving the power horse the Allis look. The tractor can pull about 80 per cent of its own weight. Instead of spinning down like other tractors, it starts to crow hop like a four-wheel pick-up.

The Power Horse was in production until 1941, when the onset of World War II in the United States forced manufacture to cease.

The Eimco Power Horse - lovingly restored by Theo McAllister of Utah. This machine was painted in brown and tan, but during the War when conditions were difficult, they were painted in whatever color was available.

Mr Bert Bonham - designer and builder of the Eimco Power Horse - seated at the reins.

Right: The instructions leaflet issued with the Power Horse shows the tractor's specifications.

POWER-HORSE SPECIFICATIONS

MODEL A-20

(1) POSITIVE DRIVE—to all four wheels, with full traction at every point of ground contact. Wheels cannot spin.

(2) TIRES—traction grip, 7.50 x 18, casings and inner tubes. Air pressure or liquid ballast. All tires alike and interchangeable.

(3) LENGTH—only 80", overall.

(4) HEIGHT—only 51", streamlined, no obstructions.

(5) GROUND CLEARANCE — 18" at lowest point.

(6) WEIGHT—2500 lbs. 2800 with ballast in tires.

(7) TURNING RADIUS—only 3 ft.—less than the width of the machine itself.

(8) TRANSMISSION—four speeds both forward and reverse. All gears mounted on anti-friction bearings running in oil.

(9) SPEEDS—(Miles per hour at 1400 R.P.M.) Low: 2.2; Second: 3.2; Third: 4.4; High: 7.9; Reverse: one-half of forward speeds.

(10) FINAL DRIVE—roller chains and sprockets, mounted on anti-friction bearings, running in a bath of oil.

(11) DRAWBAR PULL—one 16" plow, or two 12" plows, under normal conditions.

(12) ENGINE—4 cylinders; vertical; valve-in-head; $3\frac{3}{8}$" bore, $3\frac{1}{2}$" stroke; 1500 r.p.m. Diameter of crankshaft, $2\frac{1}{4}$"; number of main bearings, 3; rings per piston, 3. Removable cylinder sleeves; removable valve guides; fuel, air and oil filters; fuel used, gasoline.

(13) LUBRICATION—Engine lubrication is by controlled pressure to crankshaft bearings, camshaft and rocker arms. Control mechanism and transmission gears run in oil.

(14) COOLING—Tubular type radiator; water circulated by pump; air circulated by $14\frac{1}{2}$" cooling fan; capacity of cooling system, 2 gallons.

(15) GOVERNOR—Centrifugal type; variable speed with hand control; quick, sensitive action gives close regulation of speed and power; fully enclosed.

(16) IGNITION—Water-proof magneto with impulse coupling to assure easy starting; latest design; durable, dependable.

(17) CARBURETOR—Specially designed for fuel economy; operates efficiently at all angles.

(18) FUEL SUPPLY—Tank capacity 12 gallons; gravity feed; fuel filter.

We reserve the right to make changes in the above specifications, or to make improvements, without notice or obligation.

With this 16-inch "horse-drawn" two-way plow, the Power-Horse pulls a deep furrow in second speed at more than three miles per hour. Only a few seconds are required to unhook from the plow and hook onto a harrow.

Equal division of weight and tractive power between its four large billowy tires enables the Power-Horse to pull a harrow over newly-plowed ground with less packing down of the soil than is possible with any other type of wheeled tractor.

This "horse-drawn" hay rake never knew what it could do in a day until it got behind a Power-Horse. Another job that simply can't be done with any other type of tractor.

Eimco Power-Horse attached to "horse-drawn" cultivator making turn at end of row in a field of sugar beets. Turning space required is less than the width of the cultivator. Operator has perfect control of both Power-Horse and cultivator—can accomplish at least three times as much work in a day as with a team.

World War Two

Although tractor production slowed up in Europe, the United States export market boomed as practically every model made by the Big Nine tractor manufacturers came over the Atlantic to help with the war effort under a Lend Lease agreement. In fact, this project put more US tractors into more countries worldwide than any peacetime export drive could do.

The David Brown VIG tractor employed by the Air Ministry in Britain during the Second World War. Notice the Royal Air Force colors visible from the front.

World War Two

Agriculturally, Britain was better prepared this time, and one of the first steps that the government took was to set up the War Agricultural Committees. These committees were to oversee a plowing-up and reclamation campaign and they were also to deal with the distribution of necessary machinery. The "War Ag", as they became known to British farmers, had considerable powers and could, if needs be, evict uncooperative farmers. However, apart from the obvious grumbles about bureaucracy and controls, most farmers got straight down to the job of feeding the nation and its armed forces. The Women's Land Army was also revived to replace men called up for military service.

Tractor production in Britain was almost wholly Fordson, the N having had a near monopoly since moving to Dagenham. The only possible competition, imports apart, were the Marshall M and the recently introduced David Brown VAK-1, most others having ceased production. Fowlers who, in 1933, had begun production of a range of diesel crawlers had their factory taken over for war work and tractor production ended. David Brown managed to keep going despite materials shortages by making an airfield model for the Royal Air Force and over 100 DB-4 crawlers for the army, these being licence built Caterpillar D-4s with Dorman engines.

Everything else came across the Atlantic by convoy. Practically every model made by the Big Nine came over in varying numbers. Allis-Chalmers, who had established their own distribution

The Fowler 30 Diesel Crawler, built in Leeds, England in the late 1930s.

company in the late 1930s, brought in Us, WCs, WFs, Bs and Cs plus the odd A for threshing contractors. The M crawler was popular as were the General Motors two-stroke diesel powered HD models introduced in 1940. Case sent the D series and the LA that replaced the L in 1941, plus small numbers of R and S models. The Ds that came to Britain were the D Standard, DEX Standard (variously said to stand for export or extras), and the DC-4 row-crop. No DC-3s came over. Strangely, the SC-3 did, but not the S Standard. Both models of the R appeared in small numbers. Case had been sold in Britain since the Great War for most of the time by Associated Manufacturers of London.

Caterpillar, by now making the classic D series of 2, 4, 6, 7 and 8, found their models in use not only on farms but in reclamation, construction and coal mining as well as military use. This also happened with the Cletrac range for whom importers Blaw–Knox made bulldozers and other equipment. Many of the large crawler manufacturers found their way into industrial and military use more than agriculture.

John Deere's popular D, by now in styled form, joined the A, AR, B and BR along with a few H and L tractors for special needs. The Deere row-crops were, like most row-crops sent to Britain, used for sugar beet and potatoes, and the wide

A post-war example of the Minneapolis Moline R, that was exported from the United States during the war, and one tractor which really put the company on the map in Europe.

front axle models were preferred to the V-twin that was popular with American corn growers. Under war conditions, however, farmers got what they were given on the basis of what survived the U boat activity in the North Atlantic.

A newcomer to British farmers was the Minneapolis Moline. A few Minneapolis Molines came over previously and an importer had been established just prior to the outbreak of war. The U was imported in the largest numbers and was much liked. Zs and Rs along with GTAs helped Minneapolis Moline's reputation and after the War the company attempted to establish a factory in Britain. Minneapolis Moline offered the first factory-fitted liquified gasolineeum gas engine for tractors in 1939 with a modified U. LPG was to become a popular fuel in parts of

the United States but its use in other countries was almost nil.

Possibly the least popular range of tractors was that of Massey-Harris. Massey's new models, the 80, 100 and 200 series used mainly Continental engines with a few Chryslers. These were gasoline engines and no doubt in this form they were good but for Britain where gasoline was not used to fuel tractors, they were converted to gasoline/distillate oil. This made them very temperamental to operate, and on light work it was difficult to keep them running. Massey–Harris' main contribution was the MH21 self–propelled combine which more than made up for the shortcomings of their tractors.

Imported by John Wallace of London and Glasgow, the Oliver 70, 80 and 90

were joined by a few of the small 60 row-crops but no 99s. One rare model that came in under the Lend Lease scheme was the 80 Diesel Standard with a four-cylinder Buda engine. These were the first American tractors to use a high speed diesel engine of the type pioneered over a decade earlier in Europe. The 80 and 90 were very popular; however the 70KD suffered like the Massey–Harris tractors in being a conversion from a gasoline engine and could be equally temperamental.

International Harvester had hoped to start production in Britain and had built a factory at Doncaster, Yorkshire, but on the outbreak of war it was taken over for war production, so Britain got its IHC tractors

from the United States. The smallest Farmall was the A Cultivision, the first offset design with the engine and transmission offset to the left so the driver could see the mid-mounted toolbar. The B was similar but in tricycle style with no offset. The next two, the H and M, made up the bulk not only of production but also of exports to Britain. A few of the MD Farmall, with its gasoline start diesel engine, also came in, along with a few WD-6 and WD-9 diesels. The standard range of the W-4, 6 and 9 supplemented the Farmalls. These were all low compression engines, much more suited to British needs and were very popular and long lasting. Crawlers also came into

The ever-popular Fordson was much in use with the Women's Land Army during the war in Britain. This model from 1944 displays an attempt to make it more suitable for row-crop work by fitting narrow tires on catchpole wheels and adjustable front wheels.

International Harvester exported large numbers of Farmall models. The H (above) and the M made up the bulk of production. The B (right) was a slightly earlier wartime model built in tricycle style.

Britain, mostly diesel but some T–6 gasoline/kerosenes also came over.

Ford brought in some 9NAN tractors, but the company's main task was to keep the Ns coming off the line through bombing, blackout and endless material shortages. One by one savings were made; narrow fenders, pressed brackets instead of castings, cut away to the platform, all helped to save metal. Holes were produced in the radiator side panel to hold spark plugs for stripping and cleaning and the color changed from orange to dark green to make the machine less of a target for the German Luftwaffe. Ford, to their credit, never exploited their near monopoly and kept prices low whilst providing a good reliable and efficient tractor.

Pearl Harbor

In 1941, the Lend Lease agreement was passed, making tractor imports easier. However, the attack on Pearl Harbor in December of the same year brought Japan and America into the War, and the situation changed again.

When the Far East was overrun, rubber tires became almost unobtainable. The American tractor industry now came under regulation and, like their British counterparts, had to economize to support the War Effort. Steel wheels came back into use and even tractors like the Allis Chalmers B and the Ford 9N were produced with these. Incidentally, the 9N became the 2N in 1942 with steel wheels, hand start and magneto ignition as economy features. Designs were frozen

The Minneapolis Moline ZTX US Military tractor of 1943. With its five-speed gearbox it could achieve a top speed of 15.3mph.

Fordson Standard Land Utility Model 1938 to 1940. Registered in Lincolnshire originally, looks to be at a ploughing match or demonstration. With chains fitted to the rear wheels for extra traction.

for the duration and the Nebraska Tests suspended. A few new models did manage to emerge: the Case VAC replaced the V and John Deere superseded the G with the GM, which was styled to match the rest of the range. Deere had also added high clearance versions to most of their row-crop range in 1944. That same year Oliver bought out the Cletrac company and became the Oliver Corporation.

The German occupation of most of Europe stopped tractor production in many places. Renault managed to carry on with research into producer gas generation for tractors and the 301H of 37hp came out in 1941 along with a diesel 301D but production did not start until 1944. Renault made quite a number of prototypes with gas generators mostly carried in front of the radiator which must have made for very heavy steering and poor forward visibility. The smallest was the 25hp 304 which was approved for later production, and the largest was the 307 giving 70hp from a producer gas engine. The AFM crawler arrived with a gas conversion.

Italy, who changed sides and joined the Allies in 1943, kept small scale production going and even managed a few new models. Motormeccanica added the C50 crawler with a four-cylinder, 15hp gasoline/kerosene engine to their range to complement the S50 that had been in production since 1935. ORSI also offered a

crawler in 1942, the Super Cingoli with a
50hp semi-diesel as used on their Super
50 wheeled model. Bubba's crawler, the
42hp D42C, was on similar lines.
Production figures were low: Landini built
about 1300 of the SC50 during World War
II and about the same number of Velites.
The 40hp single-cylinder semi-diesel

Buffalo joined the line in 1941.

Allied naval power prevented Germany
importing food by sea, so part of the
conquest plans included the need to grow
more food; the wheat lands of the Ukraine
being a particular prize. Tractor
production continued from the main
companies of their previous models until

Minneapolis Moline's NTX jeep of 1944
measured just 58 inches from the ground.
The four-wheel drive achieved 36.3hp and
rode on truck tires.

the war situation forced them to stop.

Bombing and material shortages were common to both sides in Europe, however, after D–Day in 1944, battle damage worsened the situation for Germany. The chronic fuel shortage led to increased reliance on synthetic fuels, and for tractors this meant the producer gas system known in Germany as Holz Gas. Kits to convert popular tractors were made and literally attached to Hanomag, Deutz and other makes, including the "burn anything" Bulldog. Several companies made special gas models with enclosed, built in gas generators using wood, coal and anthracite burners.

Fahr's 25hp Holz Gasschlepper Typ HG25 carried the gas generator neatly enclosed on an extended chassis with the controls leading back to the driver. In Fahr's red color scheme it was a very unusual and striking machine.

Very similar models came from Normag and Fendt. The diesel did not disappear, as for small tractors the gas generation plant was too heavy to carry. An unusual tractor was the Ritscher N2O with a 22hp twin-cylinder Deutz diesel, possibly the only three-wheeled model in Germany. However, it was not a row-crop in spite of its overhead steering like the Farmall, because it had very little clearance underneath. Most tractors were well styled with front as well as rear fenders and upholstered seats and surprisingly rubber tires in many cases, although they were often small for the size of the tractor. Larger tractors came from Famo, including the Boxer crawler with a 45hp four–cylinder diesel and the 100hp Riese crawler. MAN who had returned to tractors in 1938 offered the AS250 until 1944 with a 50hp four–cylinder diesel. By late 1944 tractor production was very reduced and it ceased in 1945 for all practical purposes.

The major crawler tractor user in World War II was the Soviet Union whose army almost ran on tracks. After the loss of the Ukraine, the Kharkov plant stopped production of the SKhTZ–NATI and the fighting closed the Stalingrad factory, who also made the NATI crawler, the only model in production was the diesel-engine S-65, based on the Caterpillar 60. Turned out in huge numbers at a factory in Cheliabensk, it was supplemented by Lend-Lease crawlers from the United States. Considerable numbers of Allis-Chalmers HD7, –10 and –14 tractors along with models from Caterpillar and International Harvester arrived in Russia, mostly for military use.

The Soviet Union's Universal Y-2 tractor build in 1944 at the Vladimir tractor plant.

"Универсал У-1"
Universal У-1

"Универсал У-2"
Universal У-2

"Универсал У-3"
Universal У-3

"Универсал У-4"
Universal У-4

The Universal range of tractors from the Soviet Union. These machines featured a kerosene carburretor engine and metal wheels with grousers.

Lend Lease put American tractors into New Zealand, Australia and possibly other countries. Military use especially by engineer units, took them everywhere the war touched. In the Pacific Theatre the bulldozer was said to be "the boss of the beachhead". They were among the first equipment ashore on D-Day clearing obstacles and building roads and airstrips. A number of D-8s were waterproofed for pushing landing craft back out to sea. In his book *Muck Shifting for King George*, H Sanders tells of one bulldozer driver who had three machines destroyed by mines in one day at Solerno in Italy.

Hungarian army

Bart Vanderveen in Historic Military Vehicles Directory says that the HSCS semi-diesel remained in production and 700 went to the Hungarian Army as the KV50L followed by a number of half-track versions in 1944. These were modified with extra seats for the gun crew.

The Return of Peace

With the end of the War came a boom in tractor manufacture as many farmers - who had done relatively well from the war-time production drive - sought to replace worn-out machines. New models appeared from the major manufacturers, while many smaller companies rose up to meet the growing demand.

The Oliver Standard 77 six-cylinder tractor - brought out in 1948 as part of a new styled series.

The Return of Peace

Peace returned to Europe in May 1945 and in the Far East in September of the same year. The world's tractors were in need of repair or replacement and, just as after the Great War, the "swords to plowshares" concept came thick and fast.

The Jeep, that had become famous during the War, was offered in modified form for farm use and the following year the Empire Tractor Corporation used a Jeep engine and parts in their tractor, an idea that also occurred in Europe.

The "Big Nine" did not replace their lines immediately but added extra models. Allis Chalmers brought out their biggest crawler, the HD-19 powered by a General Motors' six-cylinder, two-stroke diesel and using a torque convertor in the transmission. Apart from a few Fordson Ns, built in Britain for aircraft towing during the war, the HD-19 was the first production tractor to be fitted with this type of transmission. Allis brought out two other innovative tractors in the rear engined G tool carrier type and the WD which replaced the WC. The WD had power adjustable rear wheel tread using spiral bars between the rims and the hubs. The tractor also had an extra clutch in the transmission that stopped the forward movement but kept the p.t.o running; however, the main clutch stopped

The war's influence carried over into farming life long after it had finished. Here a woman works a Ferguson TE20 with binder. The name TE meant Tractor England and was renowned the world over as the 'Little Gray Fergie'.

The largest of the new line Massey Harris tractors - the 55 - built in 1947.

everything running and so it was not a truly independent p.t.o.

A vertical twin powered John Deere's first post-war model when the H, LA and L were replaced by the M in 1947, followed by a crawler, the MC and the MT tricycle in 1949. Deere went into diesel production that same year with the R. True to tradition, it used a two-cylinder engine started by a pony engine. Built only in standard style it established Deere's diesel credentials.

Allis' big crawler was topped by International Harvester with the TD-24, its six-cylinder gasoline-diesel engine giving 161hp at the drawbar. Very few went on to farms, the majority being used for scraper towing and bulldozing. In about 1947, IHC fitted a hydraulic toolbar lift to

the A replacement, the Super A. The Farmall Cub also of offset design for close row-crop work was even smaller than the model A.

In 1947 Massey-Harris brought out their new range which included the well known 44 in its various configurations. Styling was similar to the previous model series and ranged from the No 11 Pony through the 20, 30 and 44 to the large 55. The 44s were Massey's most popular tractor. The engines used were based on Continental valve in head instead of the previous flat-head engine. Diesel models were available in due course on the 44 and 55, as were hydraulic lifts on all but the 55, which had slave cylinders only.

The following year, Oliver also brought out a new tractor line all styled to match.

They too had a diesel option on the 77 and 88. The 66 was a four-cylinder gasoline or kerosene/distillate, and the 77 and 88 had six-cylinder engines with the same options. All came in standard or row-crop versions and had independent p.t.o but not hydraulic lifts. The old 90/99 model continued in production until 1953 before being updated.

Harry Ferguson parted company with Ford following a disagreement. Henry Ford's son Edsel died in 1943 and Henry returned until 1945 when his grandson Henry Ford II took over. Ford were in a bit of a financial mess over the 9N/2N and Henry II's ideas did not meet with Ferguson's approval.

Ford wanted to not only build an improved model but also market it,

leaving Ferguson with only the English TE-20 to sell. Ferguson opened his own factory in Detroit in 1948 to make the TO-20 (TO - Tractor Overseas, TE - Tractor England). He then sued Ford for damages due to loss of business and patent infringement over the draught controlled hydraulic lift on the new Ford 8N. When it was finally settled in 1952, Harry drew some nine and a half million dollars.

Farmers had done reasonably well out of the wartime production drive and were seeking to replace old tractors, while farmers still using horses were wanting a tractor. Despite the relaxation of controls the major companies could not meet the rush of orders and, as happened after World War I, numerous new companies sprang up to offer tractors. Most of them

A beautifully restored Oliver Standard 88 model of 1948. Like the 77s, the 88 was available as a diesel option.

International's McCormick Super W4 tractor. The transmission offered five forward speeds and one reverse. It was rated a three-plow machine.

Independent p.t.o.

The independent p.t.o had been introduced in 1947 by the Canadian company Cockshutt Plow of Brentford, Ontario. Cockshutt had previously sold Olivers under their own name, the pre-War models having Cockshutt cast in the radiator toptanks instead of Oliver; later models were just repainted in Cockshutt's red and cream colors. The difference the p.t.o made to combine and tractor work was so great that all makes eventually had to fit independent or live p.t.o to their tractors.

Eastern Bloc countries

The Marshall Aid Programme of 1947 put many tractors into Europe when tractor production was trying to recover. Communist governments imposed by the Soviet Union on the countries under their control nationalized the farm machinery and tractor industries to supply the collective farms that were also established. One curious anomaly occurred in Hungary where Hoffer-Schrantz-Clayton-Shuttleworth kept its name and continued producing its single-cylinder semi-diesels for home and export,

some as far away as Australia. Poland produced a Bulldog copy, the Ursus, and Czechoslovakia consolidated most of the tractor industry under the Zetor name. The Zetor 25 came out in late 1945 and was a modern tractor with a 25hp two-cylinder diesel; the later ones were fitted with hydraulic lift. The Skoda 30 (above) built in 1947 at the Skoda factory in Pizen, was the other major Czech contender. Industria Aeronautica Romana, who had produced aircraft during the war, gave Romania its first tractor in late 1946. The IAR 22 on steel

wheels but with front and rear fenders on some of them, looked very much like the German Hanomag R-40. The improved IAR 23 and the larger SRT-1 followed a couple of years later. The initial color scheme was a pale orange. In East Germany, the new regime put a small tractor, the RS 03 Aktivist with a 25hp V-twin diesel into production in 1949 at Brandenburg and the similar size RS 02 Brockenhexe at Nordhousen. The large 40hp Famo diesel became the RS 01 Pionier also from Nordhousen.

A Fordson E27N model - an industrial version of the E27N Major produced from 1945 to 1951.

used bought-in parts and were assembled. Among these were Brockway, Friday, Laughlin, Earthmaster, Farmaster and the Custom, which used a Gyrol Fluid Drive torque convertor. Sheppard offered diesel tractors in three sizes and also made diesel conversions for Farmalls. Once again there was a fling at the four-wheel drive idea from Detroit who made skid steer tractors in 16, 35 and 45hp sizes and a similar tractor from the General Traction Corporation, also skid steered. The big skid steered Harris Power Horse was a much better machine with a 96hp

six-cylinder Chrysler engine, and lasted longer than the others. Silver Kings were still being made and the Co-op now sold Cockshutt tractors under the Co-op name.

Germany and France

Under the Allied occupation, West Germany continued with traditional size German farms, many of them small, especially in the south. One enterprising company, G R Wille of Hamburg, made four-wheel drive tractors out of damaged Jeeps with a diesel engine of about 15hp; like the Empire in America it only lasted a

couple of years. Damage to factories was extensive and it was about 1949 before the big companies like Lanz and Hanomag resumed production, Lanz with the Bulldog D5506 of 16hp and Hanomag with their R-25 four-cylinder diesels. These were nicely styled and painted, the

Hanomag in red and the Lanz in blue with red wheels, but as yet they lacked three point lifts.

Two of the first post-war West German tractors were four-wheel drives: the MAN 325 set the style for a line of very good tractors that lasted about 20 years until

The major producer of post-war French tractors was Renault, who built over 8500 between the end of the war and 1948. This R7012 model was produced in 1950.

the company decided to concentrate on making trucks. The other was the Boehringer Unimog, later sold to Mercedes Benz and still sold today in the same basic configuration.

Numerous small companies sprang up to make tractors. The Alpenland had a device to "enable the driver to lift the front wheels when he hits uneven ground". The others were in the main conventional small diesels. Stihl, who later made chainsaws, produced tractors from 1948 until 1963 starting with the S140 in 1948 and Faun, later to become well known for their heavy lorries, also made a few tractors in 1949/50. Others included Gutbrod– Farmax, Primus, Normag and several others through to Zanker, many using bought in diesel engines.

Fendt, who would go on to become a major tractor manufacturer, brought back their Dieselross tractors with MWM two–cylinder engines of 25hp or 16hp Deutz engines, both diesels. Fahr, Güldner, Schlüter and Allgaier who were to become part of the famous Porsche car company. Deutz, or to give them their full name Klöckner-Humoldt-Deutz-AG, would become big enough to take over Allis-Chalmers in America after swallowing up Fahr among others. Eicher tractors were noted for their use of air-cooled diesels, a feature they shared with Porsche.

The German International Harvester factory, severely damaged by bombing, built the FG12 tractor, based on the F12 Farmall, in 1948 followed by a diesel version with styled sheet metal in 1951. IHC built a No 6 producer gas tractor during the war.

The major producer in post-war France was Renault, who managed to build over 8500 tractors by 1948. Most were the 303E gasoline model until the 3042 came out in 1948; this was an excellent up-to-date tractor with a choice of engines, including a diesel. It was one of the first French tractors with hydraulic three point lift and was exported to several countries. Other French tractors coming into production included the M.A.P with an unusual horizontally opposed twin 30hp diesel and the new Vierzon 302, 401 and 402 semi-diesels. The largest was probably the Sift TL4 of 55dbhp, an impressive dark blue machine. Several small vineyard tractors were also emerging; one used the tracks from a German half-track motorcycle, the NSU Kettenkrad.

British post-war machines

The first British post-war tractor was a compromise: the Fordson E27N Major. Originally, it was conceived as a stop gap tractor in response to a Ministry of Agriculture request for an improved N capable of three bottom plowing and row-crop work. It used a slightly uprated N engine driving a three speed and reverse gearbox through a conventional clutch in place of the troublesome multiplate in the N. The rear axle was driven by bull wheel and pinion instead of the N's wormwheel. First out in early 1945, production lasted until 1951. As the post–war situation eased, an electric system and hydraulic lift became available, and also from 1948 a Perkins' diesel option became a popular alternative to the gasoline/kerosene engine.

After making 140,000 model Ns during the war, the factory machinery was very worn. Due to the lack of dollars for importing machine tools and the election of a Socialist government in 1945, Ford had to make do and it is to their credit that they did so. The Major was, as Allan

The Fordson Major tractor at work in 1948.
Its tires are ballasted with water.

The 35 was the successor to the much-loved 'Gray Fergie' from Harry Ferguson and built with Sir John Black, chairman of the Standard Motor Company, in Coventry from 1946.

Condie says in his book *Fordson and Ford*, "the bridge between ancient and modern". When in 1952 the modern came, it was well worth the waiting.

Harry Ferguson had hoped his tractor would be made by Ford at Dagenham, so when it became obvious that this would not happen he looked elsewhere. A war surplus factory in Coventry owned by the Standard Motor Company was available, so Ferguson reached an agreement with Sir John Black of Standard to manufacture a British Ferguson similar to the Ford 9NAN. Lacking a suitable engine, Ferguson persuaded the government to allow him to import the Continental Z–120 along with some new machine tools to get production started for both home sales and export. The first "gray Fergie" came off the line in late 1946. The Standard engine became available in 1947 and replaced the Continental fully that following year. The diesel option came much later in 1951. The "gray Fergie" - one of the world's best known and popular tractors - sold worldwide; it even went to the South Pole with Sir Edmund Hillary's Antarctic expedition.

Entirely new was the Nuffield, put into production in late 1948 and available in three or four wheeled versions, one of the very few British three-wheeled tractors. The Universal model was initially gasoline/kerosene only until the Perkins' P4 diesel option came out in 1950. It became a very popular tractor both at home and overseas.

David Brown temporarily improved the VAK-1 to the VAK-1a until in 1947 the first of the very popular Cropmaster models was introduced. In 1949 this became Britain's first full diesel tractor produced wholly in one factory by a single company.

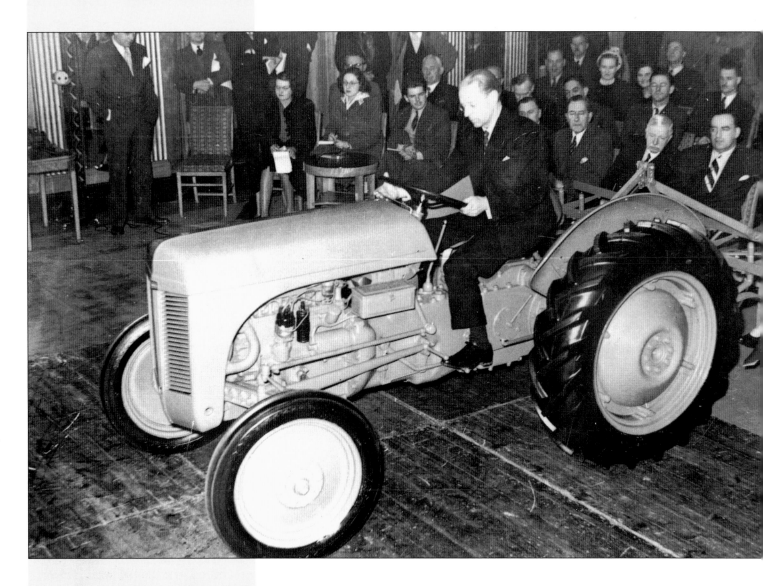

Harry Ferguson demonstrating the TE20 to a group of Soviet visitors at Claridges Hotel in London. When claims of the tractor's agility was challenged, Ferguson started the engine and drove the machine through the crowd, into the lobby and down the hotel steps. Point proven!

Between them the Major, TE-20, Nuffield and the Cropmaster persuaded British and many foreign farmers that a standard four-wheeled tractor with hydraulic three point lift and a p.t.o was exactly what they needed for their farms.

After making a few FD series diesel crawlers, Fowler of Leeds were sold to the owner of Marshall's of Gainsborough, and a crawler based on the new Field Marshall was put into production. Plans were made for a new range of crawlers using Marshall two-stroke diesels.

Marshall had revised the M and added sheet metalwork to cover the single-cylinder horizontal diesel engine. A new cold starting system used a special cartridge in the cylinder head: it should be noted that this was not a blank shotgun cartridge as some people believe.

Four American tractor makers tried to establish themselves in post-war Britain, and of these four, only International Harvester would survive. It was 1949 before IH rolled the first Farmall M out of their Doncaster plant, two years after Allis-Chalmers made a start at Tolton near Southampton. The British B used the same size engine as the American C model and late Bs also got the Perkins' P3 diesel option along with three point lift.

Massey-Harris became involved in a vast and ill-conceived scheme to grow groundnuts in East Africa, at that time still under British rule. The post-war British Government had several of these big ideas to prevent food shortages; post-war rationing was even more severe than during wartime, but all of these schemes were monumental failures. Massey-Harris needed a powerful tractor as well as one that would sell in Britain and the new 44 was found to be capable of taking the Perkins' P6 diesel in place of the gasoline engine. Using imported engineless tractors, production started at Massey's assembly plant near Manchester. However government pressure moved them to Kilmarnock in Scotland in 1949. Gradually the tractor became an all-British product but it was not a good seller, even Massey-Harris fitted three point hydraulic lift.

Minneapolis-Moline also offered a tractor for the groundnut scheme. Imported UT transmissions and rear axles were mated to 45hp Dorman or 65hp Meadows diesel engines to form the UDS

with electric starting. A p.t.o was available but there was no hydraulic lift. The tractor proved unsuccessful, especially on soft ground, and coupled with the very high price it ceased production in 1950. A few were exported. Minneapolis-Moline also tried a tie-up with the French company Mathis, and a prototype vineyard R, badged as a Mathis-Moline, was shown at the 1950 Paris Show, but nothing else came of the venture.

The desperate shortage of large crawlers after the war saw the use of converted tanks, especially Shermans with the General Motors' diesels. Several coal mining companies and farm contractors cut them down for use as tractors. Vickers managed a real conversion job with the Shervick made for jungle clearing on the groundnut scheme.

There were numerous small companies who tried the market with limited success, but most were small three-wheelers.

One, the Kendall, was announced in Farmer's Weekly in 1946 as the 'People's Tractor', cheap to buy and run. The company failed and under a new name, Newman, it lived a few years. Other very similar tractors came from Byron, Ota and possibly the best one, the BMB President. All succumbed to the Ferguson 20, especially when good secondhand ones appeared on the market.

Elsewhere, Italy kept faith with its Lanz derivatives along with multi-cylinder crawlers from Vender, Ansaldo, Lombardini and the large 70hp Breda 70D. Fiat offered the 600 wheeled tractor and 601 small crawler. The 601 and its successors were a popular model of Fiat crawler.

In newly independent Austria, Steyr returned to tractors with the 180 two-cylinder diesel in 1947 and the small 80 in 1949.

BM's two-stroke semi-diesel remained popular in Sweden into the 1950s. Volvo – the other major Swedish manufacturer later joined up with Bollinder-Munktell.

Swiss tractors included the Vevey 560, Hürliman D-100 and Bührer BD-4, all diesels of around 45hp. There were also a small number of other tractors made in Switzerland.

The Soviet Union returned the Universal to production and introduced the Stalinetz 80, a Caterpillar D-7 copy among its new crawler.

Australia's first volume production tractor came from a war surplus factory near Perth. The Chamberlain 40 with its unusual 30hp horizontally-opposed, gasoline/kerosene engine driving a nine forward speed gearbox enabled it to closely match field conditions. The Kelly-Lewis tractor was based on the 40hp Lanz that had been popular in pre-War Australia. Production problems prevented it making any real impact before the genuine Lanz returned in the early 1950s.

The Bollinder Munktell 10 tractor from Sweden, manufactured in 1947. It boasted a two-cylinder, two-stroke hot bulb engine and achieved 23hp.

The Fifties

This period of stability in the tractor market encouraged a sure and steady growth in sales and development, with not only the major players offering a wider choice, but companies and countries new to tractor manufacture coming into the ballpark for the first time.

A farmer's favorite - the Caterpillar D-2 - which was built at the beginning of the decade, but was later taken off the line to make way for the D-9.

The Fifties

he 1950s could be considered the Golden Age of the tractor, with a steady demand and a proliferation of makes and models worldwide. Countries who had never before produced a tractor joined the industry. Small scale tractor producers were numerous. In Britain, over 20 companies made tractors in the 1950s, around 45 in the United States, about 60 in Germany and even a small country like Switzerland had about twelve manufacturers. In countries that were tractor importers, for example New Zealand, the range was enormous. R H Robinson in *Farm Tractors in New Zealand* lists over 30 makes offering nearly 150 models.

A list in the *Farm Implement and Machinery Review* for June 1956 of tractors sold in Eire in 1955 is as follows: 1,898 Ferguson, 1,131 Fordson, 536 David Brown, 165 Nuffield, 146 Steyr–Daimler, 134 International, 127 Allgaier, 74 Lanz, 66 Massey-Harris, 46 Zetor, 18 Güldner, 15 Allis-Chalmers, 7 Caterpillar, Claas and Gütbrod, together with 39 Other Makes. Total 4,416.

54 makes in a small country like the Republic of Ireland; the spares and service

The Cockshutt 30 built in 1950 – the first production tractor with independent p.t.o.

The Case 400 High Crop tractor
manufactured during the 1950s.

on some must have been very
problematical.

Steady progress sums up the American
tractor industry in this decade. Tests were
revised at Nebraska in 1959 when belt
tests were replaced with p.t.o tests.
During the decade, 290 tractors were put
through the Nebraska Test, including
several foreign makes and the first 200hp
tractor the Caterpillar D-9. Horsepower
increases were also evident in wheeled
tractors as were the arrival of the big
four-wheel drive tractors. American
manufacturers also realized that the diesel
was here to stay, as was the three point
lift. The lift fitted to the British Fordson
Major, that was introduced in 1952, was

adopted as the ASAE Standard Category
Two hitch in 1959; this was basically the
three point hitch used on most British
tractors since 1945. Category One was
adopted at the same time and was
basically that used on the 1936 Ferguson
A which introduced the three point
hydraulic hitch to the tractor industry.

Allis-Chalmers introduced their Traction
Booster/SNAP coupler hitch. This used a
single point drawbar with the rockshaft
coupled to it, so that when the lift
operated it attempted to lift the drawbar,
thus providing weight transfer to the drive
wheels. One advantage of this system was
that it could be used with trailed
equipment. The implement could also be

coupled and uncoupled from the driving seat. Allis-Chalmers also offered diesel for the new WD-45. A new look and new line started in 1959 with the first of the D series, the D-14, soon followed by the D-17. This style would also be applied to Allis-Chalmers tractors built in Britain and France.

The Eagle hitch, Case's answer to the three point lift and first used on the 1949 VAC, was offered on the last models of the D and S tractors before they were swept away by the new 300 and 400 series. Case's first diesel, since the small numbers of the LH model with the Hesselman engine, arrived in 1953 with the 500. It was the larger 600 which

followed in 1957. Independent p.t.o.s were now on the options list for Case tractors along with hydraulic systems. New models with revised styling and built-in headlights were introduced in 1958 along with the new Case-O-Matic drive, first used on the 800. This drive used a torque convertor that could be locked to provide direct drive. The system never really fulfilled its potential and was eventually dropped. Case crawlers came on the market in 1957 following the purchase of the American Tractor Corporation, makers of the TerraTrac range.

Caterpillar - now the only company making farm crawlers alone - continued

In 1950, Caterpillar was the only company making farm crawlers alone, although the non-farming market claimed most of its sales. The D-2 (above) was the farmer's firm favorite.

After a hugely successful run, the 8N model was replaced by Ford's Golden Jubilee model - the NAA - in 1952. It was longer, heavier and housed a new 134ci overhead valve engine.

to improve its range although the non-farm market claimed most of its sales. The farmer's favorite, the D-2, ended and the D-9 joined the line boasting a maximum of 252dbhp in its Nebraska Test.

Cletrac crawlers, which had remained substantially the same since Oliver took them over, were replaced by a new range. This would be the last one; changes in the crawler market, competition and Oliver's sale to White Motor in 1960 finished the Cletrac crawlers completely in 1965.

Numbers replaced letters for John Deere tractors, all still using the two-cylinder engines. The 1952 tractor range, while

showing their ancestry, had been updated to include three point hitch, live p.t.o, power steering and a diesel option on row-crops as well as standard tractors. By 1957, they were again upgraded to become the 320 to 820, with engines from 29 to 75hp. There was one more upgrade to the 30 series before John Deere sprung a real surprise in 1960 with the seemingly indestructible two-cylinder tractors passing into history. A foretaste came in 1959 when the 8010 was shown; a 215hp six-cylinder diesel with articulated steering and a fully mounted eight-bottom 16 inch plow. Nothing from

The International Farmall 340 of 1958.

John Deere would ever be the same again.

After building over 422,000 examples, Ford replaced the 8N with the NAA Jubilee model in 1952. Two years later, for the first time, they offered two sizes of tractor from Detroit: the 600 and 800 series. Ford's tractor line expanded significantly and included diesel engines along with offset, high clearance and tricycle row-crops.

Massey Ferguson

In 1953 Harry Ferguson sold his tractor company to Massey-Harris, this merger being the subject of numerous books, and suffice to say his age and a growing interest in cars were among the reasons for his decision. Massey-Harris found itself having to run a two line policy until everything was sorted out.

The small Massey-Harris tractors had undergone numerous changes in an attempt to enter the Ferguson- and Ford-dominated end of the market. They were dropped and replaced by the MH50 and the mechanically identical Ferguson 40 which sported different sheet metalwork. Both tractors were based on the Ferguson 35 that came with the takeover. The 40 and 50 had the tricycle row-crop option

Originally the Oliver Fleetline - the Super series was launched in 1954 with more power. Shown here is the Super 77 model built in 1957.

that the 35 lacked. The final Massey-Harris tractors were the 333, 444 and 555, which were updated versions of the 33, 44 and 55 respectively.

In early 1958, the company was renamed Massey Ferguson. The new tractors were based on the Ferguson type of tractor, the first ones being the Massey Ferguson 65 and 35. These housed the Perkins' diesel engines, and Massey Ferguson bought Perkins in 1959.

The wartime International Harvester tractors had been uprated as the Super series pending the new numbered models. These, although similar in appearance to the previous tractors, offered TA torque

amplifier transmission on some models, TA having been pioneered by the Farmall Super MTA in 1954. The Fast Hitch, a two point lift which IH hoped would rival the three point lift, was also available along with independent p.t.o, by now coming into use with most tractor makers.

Minneapolis-Moline bought out B F Avery and incorporated the Avery tractor into the Minneapolis-Moline range. The new 335 and 445 tractors incorporated hydraulic three point lift and the large models had diesel options. One of the last of the U tractors had been MM's first American diesel in 1952. The GB diesel, with over 60hp, was a pointer to

Two 1950s machines from Minneapolis Moline. Above: the ZA, produced between 1949 and 1953 it achieved 36.2hp. Left: The G900, a later model which continued to be built into the 1960s shown here with an LPG engine.

the big diesel tractors that farmers were demanding as farm sizes started to grow. Minneapolis Moline's version of the torque amplifier, Ampli-Torc, was available on the 335 and 455, and their replacements, the Star series.

The Fleetline from Oliver became the Super series in 1954 with more power. Top of the range, the Super 99 had a choice of engines, Oliver's own six-cylinder diesel or a General Motors' three-cylinder, two-stroke diesel, and when the next update came in 1958 the choice was wider still. The 950 had a six–cylinder engine burning either gasoline or diesel, the 990 used a version of the General Motors' diesel used in the Super 99 and the 995 Lugmatic was the same as the 990 only fitted with torque convertor drive. Oliver offered a small offset tractor during this period, the

Super 44, later 440, as well as their first utility tractor with built in three point hitch, the Super 55, in 1954.

Utility tractors of the type popular in Britain and Europe were starting to appear in American tractor companies' model lines. International Harvester had brought out their 300 as a utility model alongside the 300 Farmall. The small Case, Deere and other makes also showed the utility influence.

In contrast to these, the market for very big tractors started to emerge. The standard top of the range from all makers by the late 1950s over 70hp; the Minneapolis–Moline GVI and the International Harvester 660 diesels were 78hp and the Case 930 was 80hp: this represents a doubling in horsepower over two decades.

The John Deere 4020 Diesel with its differential lock to overcome problems of excessive wheelslip.

Two makes that were to become well known for giant multi-wheel drive tractors, Steiger and Wagner, started in the 1950s. Wagner started in 1953 with a medium sized model and by 1959 the TR14A tested at over 155hp. Steiger tractors originated from a home built tractor on the Steiger brother's farm and they too had a range of models in 100 to 200+hp sizes. First of the major manufacturers was John Deere with the 8010 turning out 215hp from its General Motors six-cylinder, two-stroke diesel. In 1960 Deere brought out the all new four- and six-cylinder models, the 1010, 2010, 3010 and 4010. These tractors would set a standard for many years to come. One unusual achievement of the 4010 model sold in Britain was that some would be fitted with spade lug rear wheels to overcome traction problems. The last spade lug tractors, the 4020, had a differential lock to overcome problems of excessive wheelslip.

Cockshutt extended their range and some used British Perkins' diesels. They also supplied Co-op with tractors to sell under the Co-op name and color scheme. Gambles Stores also sold the Cockshutt 30 as the Farmcrest 30.

European tractors also showed an upward power trend although not on the American scale. Farms considered to be quite large in Europe were small by North American standards so the starting point for tractor horsepower was lower, at around 25hp.

An interesting make to come on to the British market was the Turner Yeoman of England in 1951. Power came from a 40hp

Built in Britain - International Harvester's Farmall BMD. It was basically the BM model with direct start diesel.

V-4 diesel of Turner's own make and based on a marine engine. Unfortunately, although it was a well designed tractor and pleasant to drive with an upholstered bucket seat, it suffered from mechanical failures, and that, plus the high price stopped production in 1957. This tractor was one of the first to use a dry aircleaner although this was not successful and was soon replaced by an oil bath type. Dry air cleaners would become almost standard in later years.

Massey–Harris modified the 744 to use a Perkins L4 engine and produced the 745 until the new Massey Ferguson tractor came out. Following their takeover of Ferguson in America, Massey–Harris had to come to agreement with the Standard Motor Company who built the Ferguson in Britain and France. After a lot of negotiation this was finally achieved and

the gray and bronze FE35 became the red and gray MF35 The Standard four–cylinder diesel was soon replaced by a Perkins three–cylinder.

International Harvester put their first direct start diesel into the BM to give the BMD. This became the Super BMD and was joined in 1954 by the Super BWD-6 similar to the American model but with three point lift. IH also bought the old Jowett car factory at Bradford, Yorkshire and introduced one of their most popular tractors the B-250 in 1956 which was joined two years later by the B-275 and B-450 which replaced the SBWD-6. For some reason the B-450 never had live p.t.o which limited its uses, but nevertheless it stayed in production until 1970. The first British crawler, the BTD-6, came in 1953 and after uprating to 50hp in 1955 enjoyed 20 years of production.

Allis-Chalmers replaced the model B in 1955 with the D270, seen here at work. This model lasted only two years before it was superseded by the D272.

After a long run, the Field Marshall came to an end in 1957. The Series IIIa, now orange instead of green, and with a six speed gearbox, hydraulic lift and central p.t.o had become outdated and outclassed. Marshall had a go at the big tractor market mainly with a view to exports with the MP6 using a 70hp Leyland diesel; basically a good tractor but it was lacking hydraulics of any sort and it did not sell, only 197 being made before production ceased.

They were not the only company to do this: David Brown, who had gained a sound share of the market with their diesel or gasoline/kerosene 25 and 30 tractors, offered the 50D in 1953. This six-cylinder, 50hp model lasted until 1958 during which time some 1260 were sold. David Brown also produced crawlers based on wheeled tractors at this time, these machine using Cletrac style steering.

Fowler's rival to imported American crawlers, the Challenger, got off to a bad start: the Marshall two-stroke diesels proved to be very unreliable. Re-engined with Leyland and Meadows' units they sold quite well for a while but when Caterpillar opened their factory in Scotland they faded out.

Vickers also tried to get into the big crawler market with the Vigor and Vikon crawlers. These used Rolls-Royce diesels and unusual tracks with sprung idlers along the full track length rather than top and bottom rollers, but they too eventually lost out to Caterpillar. Both Vickers and Fowler supplied tractors to the British Army engineering units.

Crawlers enjoyed a spell of popularity during the 1950s and several makes made brief appearances but only one, the Track Marshall series, became popular and still remains so in a much updated style.

Of the other British tractors, the very popular Nuffield was joined by a small three-cylinder diesel, the Universal Three and Allis-Chalmers replaced the B with the D270 in 1955 which was in turn replaced by the D272 in 1957. The row crop tractor market was rapidly disappearing in Britain and Allis-Chalmers' last British tractor, the ED40 built from 1960 to 1968 was more of a general purpose design. Allis closed down in Britain in 1971.

Big tractor news was Ford's unveiling of the new Fordson Major in 1952. One of the most popular tractors from Ford, it became the basis for countless conversions. Whereas most diesels had a reputation for being bad starters, the Major earned a reputation for always starting, no matter what the conditions, and this soon ended the production of the optional spark-ignition engines. In late 1957, the Dexta came out with draught control, hydraulics and a three-cylinder Perkins based engine. Together with the later Super Dexta they were improved and uprated until the World Range came out in 1964.

County who had made crawler versions of the E27N soon offered the same for the new Major and later also made four-wheel drive conversions. Roadless Traction had also made E27N based crawlers and after making a few crawlers based on the new Major, concentrated four-wheel drive conversions.

Both County and Roadless Traction went on to become large scale makers of Ford based four-wheel drive tractors and only went out of business when Ford strated to make their own four-wheel drive tractors 30 years later.

Introduced in 1957, the Nuffield Universal Three model used a three-cylinder BMC diesel engine developing 28hp. This model boasts the 'Sta Dri' cab.

RENAULT D 22 1956

Well-suited to French farming needs - the D22 Renault tractor was launched in 1956 following the creation by the public authorities of 'improved domestic fuel oil' - a high quality diesel. These models offered diesel engines, air or water cooling, differential lock, power take-off and an hydraulic lift system.

French wines

The selection of French tractors almost rivalled the number of French wines in the 1950s, and quite a lot of the tractors were aimed at grape growers apart from the specialist machines most of the smaller standard tractors had a narrow vineyard version. At the beginning of October 1954 eight countries exhibited tractors at the Rambouillet, Seine et Aise. Seventy-two different makers used 400 tractors in the four-day trials.

Renault entered the 3040 series with a choice of engines: 22–30hp gasoline, 18–25hp gasoline/alcohol or Perkins P4 diesel of 22–30hp. The four-wheel drive included the latest in a long line of Latils and some new ones. The Agrip tractors had equal wheels and kingpin steering on

both axles. The single cylinder opposed piston two–stroke engine running at1500rpm on domestic fuel oil was fitted to the smallest model which came out in 1951.

Capable of working in either direction was the skid–steered BM 1200 from Hy. Bergerat, Monnoyeur & Cie. A six speed fully reversible gearbox along with seat and controls enabled the tractor to work in either direction to suit the job in hand. There were several other four-wheel drives available as well, for example the Buffle, as well as crawlers from Continental and the St Chammond PM5.

More conventional mechanics came from Vendeuvre who would become the French Allis–Chalmers branch à few years

The 1950 Swiss Meili tractor built under licence in France.

later. AC France made the FD–3,4, and 5 tractors for a few years in the 1960s, styled to match the American D models. Around the same time Case took over another exhibitor, Vierzon, who made two-stroke semi-diesels of an unusual design. After the Case takeover, several of these tractors came out with the Case name on them. The French Cases were the CF250 and CF350 of the early sixties.

Not entirely successful in America but popular in France, was the Massey–Harris Pony 812. A Peugeot gasoline engine replaced the Continental and later a Hanomag two–cylinder diesel came out in the Pony 820. After the merger it was restyled as the MF21 until the new unified range came out.

Purely French tractors included the Champion Elan range using Ceres

A Deutz air-cooled diesel - model 612.

two-cylinder diesels in four sizes from 24 to 36hp driving through a seven forward and two reverse gearbox. P.t.o, pulley, adjustable track width and differential lock, which was now becoming a common extra, were among the options available.

Peugeot engines were a popular choice in gasoline tractors. L. Babiole used one in the Multi Babi 203, as did ETS Semial in the Moto-Standard-Semial 54.

French worries over fuel showed up again in 1959 when Labourier, one of the larger French tractor manufacturers exhibited a polycarburant engine of 55hp in one of their tractors. This was claimed to run on anything from alcohol to Brilliantine hair oil! Too bad if you were a bald farmer.

The diversity of France was mirrored in Germany where the choice was equally wide. By the mid 1950s, major makers were offering a range of good modern tractors. Deutz changed over to air-cooled diesels and the lack of a radiator gave these green tractors with their round-fronted hoods a distinctive appearance. The smallest F1L 514/40 was a single-cylinder machine and the largest 60hp F4L 514 had four cylinders with a two-cylinder 28hp and three-cylinder 42hp completing the line. The Bulldogs remained much the same with the smaller models getting hydraulic lift.

Hanomag stayed with the water-cooling and the R range in red or blue were in two styles. The smaller R-12 and R-24 had a row-crop style with streamlined hoods while the R-45 and R-55 had retained a more traditional look. Hanomag made tractors until 1971 when changes in the market and ownership led them to a change of policy, although industrial crawler production continued under several successive owners.

Made in Wurtenberg, Germany in the early 1950s, the Porsche Allgaier three-cylinder diesel. Professor Ferdinand Porsche had been involved in tractor design since 1938.

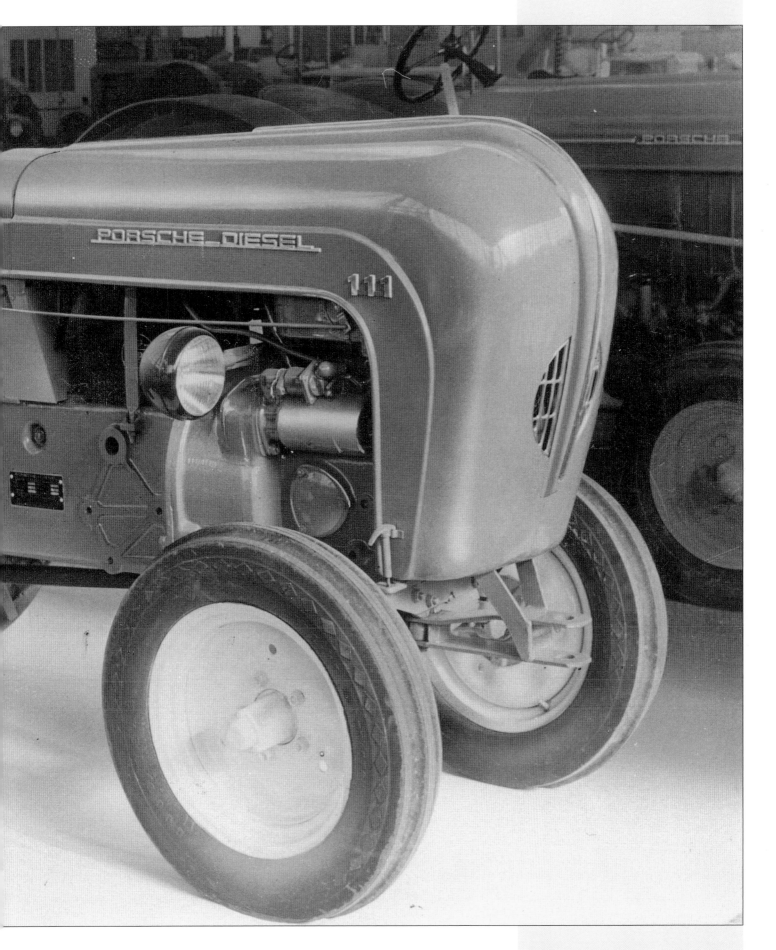

Eicher, Fendt and Fahr were also making similar ranges, Eicher like Deutz favoring air-cooling. By now practically all tractors were diesels and the four-stroke gradually took over from the two-stroke.

MAN extended its four-wheel drive range and Normag, Güldner, Hatz and many others produced tractors of usually under 40hp. Porsche took over Allgaier and made a unified range with a lot of common parts until 1964 when the company had a change of policy. The largest models had fluid flywheel clutches, draught control, multiple p.t.os (most continental tractors have a mid-point p.t.o underneath for the mid-mounted mower favored on the European Continent) and differential locks.

The German and French International Harvester factories were starting to produce an array of tractors under the Farmall name but on more European style wide front axles. The French factory made the Cub, the FC and the FCC based on their American equivalents. The German factory produced its own designs from the 14hp DLD-2 to the 30hp D430S

German tractor sales were approaching saturation point by the early 1960s and many companies, big and small, either closed or were taken over.

A bewildering choice faced the Italian farmer as over 50 home produced makes as well as imports from all over Europe and a few American tractors were available for him to choose from. Landini upgraded their semi-diesels and brought out their last model in 1957, the 44 Major. The end was in sight for the semi-diesel and Landini's new tractors

Workers at the Bollinder-Munktell factory in Sweden.

were full diesels with vertical engines, and at the top of the range licence. built Perkins' diesels. Landini was for a while part of Massey Ferguson as their crawler suppliers.

Lamboughini were a recent arrival that were to become a major Italian tractor manufacturer who would later join up with Same and the Swiss Hürlimann company. They bought engines from MWM in Germany and the popular Perkins models from Britain.

It sold both wheeled and crawler models, many of which were exported from the small La Piccola with its high waist for cultivation to the big 55hp 55R Fiat had a model for most conditions. Destined to become a major force in tractors, Same made a modest start in 1950 with the 4R20M with a 2hp, two-cylinder gasoline/kerosene engine. The next models were all diesel; starting with the DA25 the range and size grew to the 62hp DA630 by 1960. Four-wheel drive for which Same would become well known appeared on the DA25DT in 1954.

The Low Countries

To satisfy local demand a small number of tractors had been built in Holland by Brons in the 1940s and in Denmark by Bukh. Belgium too had a tractor in the 1950s, the Galman. Austrian production remained with Steyr whose range increased from single to multi-cylinder diesels with the three-cylinder 185 and the four-cylinder 280. Lindner was the other Austrian tractor.

Apart from the main three, Swiss production included some special models for the mountain farmer: the Plumett from Plumettaz & Cie had a built in winch at the center of gravity as implements were often winched up and down steep slopes. Powered by a Ford 14hp engine driving through a three forward and reverse gearbox with the option of doubling them up. Big for Swiss tractors were the Hercule 35 and 60hp models. The same company also made a Hispano-Suiza diesel powered model for sale in France.

Zadrugar, the first Yugoslav tractor came out with a Perkins P4. Yugoslavian tractors became IMT and were based on Massey Ferguson designs.

Licensed production or foreign owned factories made up most of the Spanish production. One odd result was that in 1964, when the Super Major ceased in Britain, the plant was sent to Spain and after a complex series of ownerships the Ebro tractor became a mixture of Ford and Massey Ferguson.

As the Soviet grip on Eastern Europe tightened, the individual national companies ranges were phased out in favor of a unified tractor. Zetor from Czechoslovakia, Polish Ursus and the East German ZT became almost identical as a result of this unification. HSCS survived until the late 1950s when it became the Dutra (dumper and tractor) works. In a complete departure from other models the new D4K was an equal wheel four-wheel drive of nearly 100hp. Bulgaria started production of the small Bolgar crawler tractor. The Romanian tractors were renamed the Universal and used Fiat tractors as their design basis for the smaller models. Some were sold in the United States as Long tractors.

The Soviet Union love of crawlers started to wane as tractor designers discovered the rubber tyre: most wheeled tractors had been on steel wheels as tarmac roads were a rarity. The new factories specialized in one type of tractor,

Road travel

The popularity of the small crawler in Europe led to several attempts to make it capable of travelling along roads to and from work. ITMA in Italy offered one of these systems: ordinary type tractor rear tyres were bolted to the drive sprockets and a front axle, almost like an early Farmall wide axle, complete with steering column over the engine, was fitted to the front. Quite a stirring vehicle to drive down the highway with the tracks and wheels all in motion. There were also tractors that could be converted from wheels to tracks and visa versa in about four hours. How many of these were made is not recorded but they were certainly advertised.

The Finnish Valmet entered production with a 13hp tractor before offering a very rugged tractor for Arctic farming where farmers go logging in the winter. Years later Valmet went to the other climate extreme and opened a factory in Brazil with a larger production than that of the original one.

Bollinder-Munktell abandoned the old hot-bulb engine with the new BM230, BM35 and BM55 diesels with direct injection, five speed gearboxes, live hydraulic lift and p.t.o, with the option of a cab: Sweden would require safety cabs in the near future. Volvo opted for a Perkins L4 in their T-30 tractor. BM and Volvo would later join forces.

the Belarus range came from Minsk and started with the MTZ-2 diesel in 1953. These were modern style tractors with bench seats and hydraulic lifts. The Chinese built a similar tractor called the Iron Buffalo as well as a crawler, the Tung-Fang-Hung, based on the DT-54. Tractor production expanded in the Soviet Union and a whole range of tractors covering every crop and job came from the factories in huge numbers. They also started to export, a far cry from the big importer they were in the 1930s.

In Australia, Chamberlain expanded their line of diesels with Perkins engines in the Champion. One of these took part in the round Australia car rally after being modified to increase top speed to 65mph. An advertisement in *The Cultivator* for September 19, 1959 gives an account of the epic journey of 11,140 miles in 19 days. 20 disabled competition cars were towed, one for 300 miles. Tail End Charlie averaged 50mph on some sections. Chamberlain was eventually sold to John Deere.

The Russian Unversal tractor at a promotion in 1952 in Moscow.

The 1960s Onwards

With the huge strides made in agriculture generally, and the vast changes in farming methods, it is little wonder that tractor design should escape unscathed. Together with the new technology of computerization and advanced communications, tractor designers and builders look towards the next century in anticipation and interest.

The highly versatile JCB Fastrac is
capable of speeds up to 50mph on
normal roads - hence its name.
Here, it is spreading fertilizer.

The 1960s Onwards

Forty years earlier had seen a tractor boom and the current one seemed set to continue. A revolution was quietly taking place as new technology became readily available. Turbo chargers used the exhaust gases to pump more air into the engine and helped boost the power from a given size. This in turn required dry air cleaners, oil coolers and other improvements. Combined, these features made the diesel the supreme tractor engine.

To make use of all this extra power, the transmissions used improved change on the move syncromesh gears and clutchless powershift became standard on many of the larger tractors. Hydraulic drive with its stepless speed changes had long been a tractor designer's dream. Roadless Traction tried it in a 6/4 Ploughmaster and Eicher in an HR one-pedal control tractor in 1964. International Harvetser put it into production with the 656 in 1967. Hydrostatic drive in practice did not come

up to expectations and was eventually scrapped. P.t.o speed went up to 1000 rpm to power bigger machinery and the pulley disappeared. Wider and multiple tires put all this power to work. Four-wheel drive in both equal and unequal wheel styles came from manufacturers rather than the specialist companies. Mechanical linkage to steering was supplemented by hydrostatic and hydraulic assistance to brakes, and gear changes eased the driver's load.

The Minneapolis Moline G1050 from 1970 with its 504ci engine.

The Allis-Chalmers D19 tractor manufactured in 1962.

Drivers, too, received more consideration than they had done in the past; sprung seats and the occasional shock absorber being replaced with fully adjustable and upholstered seats, the design of which was assisted by medical experts. Cabs had been fitted to early tractors and then gone out of use. They returned with soundproofing and often air conditioning.

Horsepower continued to grow. 100hp tractors becoming common and the 200+hp models not unusual in many parts of the world. Signs of this had been seen in the late 1950s when farmers began coupling tractors in tandem to get more power. In Britain, Doe did this on a commercial basis with the Triple D using two Fordsons to provide 100hp and

four-wheel drive. Doe also built implements to match. Major manufacturers - especially those in the United States - started to add big four-wheel drives to their line; the IH 4300 and Case Traction King 1200 were early examples. Schlüter in Germany brought out the 130hp S1500V in 1966 and similar tractors appeared in the Soviet Union, notably the Kirov series.

During the 1960s the multinational tractor companies started to rationalize their models. Both Ford and Massey Ferguson announced new lines in the middle of the decade. Ford introduced the all new 6X range in November 1964. In Britain they also dropped Fordson and became just Ford when they opened their new factory at Basildon, Essex. New to

Britain was the Selectospeed clutchless gearbox first introduced in America in 1959 and now offered as an option on all models except the small 2000. Although popular in the USA, Selectospeed did not catch on elsewhere and when the range was revised it was quietly dropped and replaced by the Dual Power system.

Ford expanded its range in the 100hp plus class with the 8000 and 9000 models in 1968. For the next 20 years, Ford tractors were based on these new models.

Massey Ferguson's Red Giants that followed in early 1965 were less radical in their design and they too soon topped the 100hp mark. They proved to be a very popular tractor line.

International took rather longer. Not until 1971 did they unify their range on a worldwide basis.

The other American manufacturers who catered mainly for the domestic market

The G706 four-wheel drive from Minneapolis Moline and manufactured in the 1960s.

A close look (above) at the Oliver 1955 (and opposite), built in 1972.

continued to expand and improve their product lines. A hint of the future direction of the industry came with the purchase of Oliver by the White Motor Company who went on to buy Minneapolis Moline and Cockshutt. White eventually amalgamated all three as Whites with a new color scheme of silver and gray.

A new feature of the American tractor market was the appearance of foreign makes, initially British, with David Brown, who took over some of the distributors dropped by Ford when they reorganized, and Nuffield offered by the Frick Company of Waynsboro, Pennsylvania. The German Deutz soon followed as well as the first Japanese tractors from Kubota and Satoh.

Oliver also started a trend that would become widespread in the future, selling foreign tractors under their own name. Initially they bought David Browns from Britain and then Fiats from Italy to fill out the lower end of their range. When Tenneco took over Case and then David Brown in Britain, the Browns replaced the smaller Case models and with the subsequent acquisition of International Harvester, the whole model system changed again with the David Brown range disappearing altogether.

The old time prairie tractor was reincarnated in the 1960s when very big four wheel drive tractors became popular.

Farmers have always built what they needed if they could not buy it and in the post-war period some American farmers decided that the crawler needed replacing. The Wagner brothers were one of the first offering hinge-steered tractors of over 200hp in later models. Shortly after a change of ownership they ceased production in the late 60s.

Steiger who also started on a farm in the 50s is still in production after several changes of ownership and periods of

The recession of the 1980s and the shrinking of the world tractor market led to Ford and Fiat joining up, with the New Holland name (long seen on combines) appearing on both the former Ford and Fiat products.

The Allis-Chalmers D19 (above), with its 42.6ci engine (right) manufactured in 1962.

building for other companies including Allis-Chalmers and International.

In Canada the Versatile Company from Winnipeg offered hinge-steered tractors from 1966 and built up a good reputation both in North America and abroad.

Larger tractors still came from a few specialist builders who made models of up to 600hp in the case of the Rome 600C.

As farm sizes increase, there will always be a market for the monster machine. In Britain, after various updates, the Nuffield range was replaced by the Leyland tractors from a new factory at Bathgate in Scotland. They did not sell as well as expected and after a change of ownership and renamed Marshall, production ceased in the 1980s.

David Brown's fate has already been noted leaving Track Marshall as the only wholly British tractor company at the time of writing. Once Britain had joined the European community, imported tractors appeared in increasing numbers and only the big three: Ford, Massey Ferguson and Case I.H. were able to make and sell tractors in Britain. When Ford started four wheel drive production, the two main specialist companies, Roadless and County, were put out of business – a fate shared by Muir Hill – who used Ford components in their tractors.

The European tractor population reached saturation point in the 1960s and manufacturers started on the road to amalgamation. These enlarged companies

ALLIS-CHALMERS

looked to exports worldwide to increase sales. Licences were also granted or they set up their own factories abroad.

Deutz tractors were built in South America and they moved into North America via a share in Steiger. In 1985 they bought Allis-Chalmers to form Deutz-Allis.

European tractor company ownership became even more complex in later years. The recession of the 1980s and the shrinking of the world tractor market led to Ford and Fiat joining up, with the New Holland name (long seen on combines) appearing on the former Ford and Fiat products.

The major names are now John Deere, the Deutz group, Same group, Case I.H.,

Massey Ferguson along with Fendt, Eicher and a few other specialist companies. In Scandinavia, Volvo and Valmet co-operate, while Renault dominates French production. Italy alone seems able to sustain more companies than its size warrants in today's climate.

The big unknown at the time of writing is the future of tractor production in the former Soviet Union and the East European countries.

Large scale tractor production also ceased in Australia when International Harvester closed down followed by the Deere owned Chamberlin plant in the mid-1980s. One aspect of Australian production was the large special order tractor mainly for the wheat farmers.

Renault's Super series introduced in the 1960s. Shown at work is the Super 5D tractor in 1965.

The Upton HT 141350 with its 350hp engine was claimed to be the world's largest two wheel drive tractor. However the 1980s recession saw this and the others suspended pending better times, leaving Australia without a native tractor.

We are currently in the electronic and computer age and inevitably these have been added to the tractor accessories just as hydraulics had been 50 years earlier. Engine control has become engine management and onboard computers adjust the hydraulic lift and transmission. Data can be collected and at the end of the job, transferred to the farm office computer for study.

Crystal ball gazing is a notoriously unreliable occupation and in looking into the future of the tractor it is worth looking back at past predictions. People were sceptical of the rubber tire and the diesel engine, both now universal. There have been many predictions of the end of oil supplies, notably after the 1974 Middle East crisis. Many attempts have been made to find alternative fuels. Allis Chalmers built an electric tractor with fuel cells to provide the power and a lot of effort, in Britain and the Soviet Union,

An experimental tractor - the HT340 from International Harvester - with its gas turbine, was tested out in the 1960s.

went into tractors using mains electricity. Alcohol from crops and bio-diesel are current future fuel predictions and will no doubt come some day.

A report in the *Implement and Tractor* magazine in 1963 expected gas turbines and hydrostatic transmissions to be the most likely future developments. Hydrostatic has been and gone and the gas turbines did not materialize except on the I.H. HT340 experimental tractor.

So what does the future hold? Consideration of needs other than that of the farmers could be one aspect. We have already had increased safety enforced via safety cabs, at least in Europe, and this has resulted in the modern cab with its excellent driving conditions. Environmental regulations on exhaust emissions will have more effect on tractor engine designs in the future. The increased highway use and the need to cope with other traffic will also influence design. The JCB Fastrac is a product of this.

The satellite positioning now in use on combines to record crop yields and thus fertilizer application could possibly be extended to field work and the amount of cultivation needed in the field. With the larger fields, the variation in soil type while working across on any job is increasing and an accurate soil map fed into the tractor computer could change speed and seed rates, for example, without the driver's involvement. The electronic possibilities seem limitless. On a more mundane level it is more likely that the current tractor will continue very much as it is at present, considering that the numerous attempts to redesign the basic layout of Henry Ford's "F" in 1917 have never lasted for long. We shall see.

The 1,000hp tractor? Well, that was the Japanese Komatsu D555A crawler in 1982.

Power-assisted steering, four-wheel drive and improved gearbox design were all incorporated into the Renault models of the mid-1970s.

INDEX

FURTHER READING

Ackergiganten by Klau Herrmann, Georg Westerman Verlag GmbH, Braunschweig, Germany, 1991

Agricultural Tractors - their Evolution in Australia by Ian M Johnston, Kangaroo Press, Kenthurst, NSW, Australia, 1995

The Allis-Chalmers Story by Charles Wendel, Crestline Publishing Co, Sarasota, Florida, USA, 1988

The American Farm Tractor by Randy Leffingwell, Motorbooks International, Osceola, Wisconsin, USA, 1991

Australian Tractors by Graeme R Quick, The Land Book Company, Richmond, NSW, Australia, 1990

Barbed Wire Tightener to Power Plus by Vannu Niskanew, Valmet Tractor Corp, Jyvaskyla, Finland, 1989

Caterpillar by Randy Leffingwell, Motorbooks International, Osceola, Wisconsin, USA, 1994

Caterpillar Tractors 1926-1959 by Peter J Longfoot, P J Longfoot, Peterborough, Cambridgeshire, UK, 1993

Classic American Farm Tractors by Nick Baldwin, Osprey Publishing, London, UK, 1985

Classic Tractor Special 1-3 by Allan T Condie, Allan T Condie Publications, Carlton, Nuneaton, UK

Classic Tractors in Australia by Ian M Johnston, Kangaroo Press, Kenthurst, NSW, Australia, 1993

Encyclopedia of American Farm Tractors by Charles Wendel, Crestline Publishing Co, Sarasota, Florida , USA, 1979

Endless Tracks in the Woods by James A Young and Jerry D Budy, Crestline Publishing Co, Sarasota, Florida, USA, 1989

Farm Crawlers by R N Pripps and Andrew Morland, Motorbooks International, Osceola, Wisconsin, USA, 1994

Farm Tractors (3 vols) by R B Gray, Lester Larsen, Larry Gay respectively, American Society of Agricultural Engineers, St Joseph, Michigan, USA

Farm Tractors in New Zealand by Richard H Robinson, Country Life (NZ) Ltd, Ngongotaha, New Zealand, 1989

Fifty Years of Garden Machinery by Brian Bell, Farm Press, Ipswich, UK

Ford and Fordson Tractors by Michael Williams, Farm Press, Ipswich, UK

Ford Tractors by R N Pripps and Andrew Morland, Motorbooks International, Osceola, Wisconsin, USA, 1991

Frank Munktells till Valmet by Olov Hedell, Media Nova, Stockholm, Sweden, 1994

Great Farm Tractors by Charles Wendel, Bison Books, London, UK, 1995

Hanomag Schlepper by Armin Bauer, Franckh W Keller & Co, Stuttgart, Germany, 1989

Implement and Tractor 100 Years edited by Robert K Mills, Intertec Publishing Corp, Overlark, Kansas, USA, 1986

John Deere Farm Tractors by Randy Leffingwell, Motorbooks International, Osceola, Wisconsin, USA, 1993

John Deere Tractors and Equipment (3 vols) by Don Macmilan and Roy Harrington, American Society of Agricultural Engineers, St Joseph, Michigan, USA

John Deere Two Cylinder Tractors by Michael Williams, Farm Press, Ipswich, UK

Lanz Bulldog 1928-55, Schrader-Motor Chronik, Munich, Germany

Massey Ferguson Tractors by Michael Williams, Farm Press, Ipswich, UK

Minneapolis Moline Tractors 1870-1969 by Charles Wendel and Andrew Morland, Motorbooks International, Osceola, Wisconsin, 1990

Minsk Tractor Plant published by MT Plant, Minsk, Russia

Modern American Farm Tractors by Andrew Morland, Motorbooks International, Osceola, Wisconsin, 1994

Nebraska Tractor Tests since 1920 by Charles Wendel, Crestline PUblishing Co, Sarasota, Florida, 1985

1918-1988 Renault 70 ans de Tracteurs Agricoles, Renault Agriculture, France, 1988

Oliver Hart Parrr by Charles Wendel, Motorbooks International, Osceola, Wisconsin, 1993

150 Years of International Harvester by Charles Wendel, Crestline PUblishing Co, Sarasota, Florida, USA, 1991

Quattro Ruote che lavorano by Massimo di Nola, Edizioni del Sol, Milano, Italy, 1987

Schlepper by Armin Bauer, Franckh W Keller & Co, Stuttgart, Germany, 1987

Schlepper Klassiker (2 vols) by Jurgen Hummel, Franckh-Kosmos Verlags GmbH, Stuttgart, Germany

Taschenbuch Deutsche Schlepperbau by Wolfgang H Gebhardt, Franckh W Keller & Co, Stuttgart, Germany, 1988

Tracteurs Agricoles en France (1910-1960) by Claude Ampillac et Bernard Galvat, Editions E/P/A, France, 1993

Tractor Album Series: Ferguson, JCB, Allis-Chalmers, Massey Harris, Nuffield, Marshall Fowler by Allan T Condie, Allan T Condie Publications, Carlton, Nuneaton, UK

Tractor Heritage by Duncan Wherret with Trevor Innes, Osprey Automotive c/o Reed Consumer Books, London, UK

Tractors at Work (vol 1 and 2) by Stuart Gibbard, Farm Press, Ipswich, UK

Tractors since 1889 by Michael Williams, Farm Press, Ipswich, UK

Tracktoren in Deutschland 1907 bis Heute by Klaus Herrmann, DLG-Veerlags GmbH, Frankfurt am Main, Germany

Veterenen der Scholle by Armin Bauer, Landwirtschafsverlay GmbH, Munster, Germany

Vintage Tractor Special series 1 to 17 by Allan T Condie, Allan T Condie Publications, Carlton, Nuneaton, UK

World Harvesters by Bill Huxley, Farm Press, Ipswich, UK, 1995

ACKNOWLEDGMENTS

The author and publishers would like to thank the following people and companies for their help in compiling this book.

Jonathan Brown
Lydie Chalumeau
Alan Condie
Richard L Davidson
Peter Dean
Ted Everett
David Fletcher

Ford Motor Company
Hart-Parr/Oliver Collectors Association, USA
Bil Huxley
Massey Ferguson Tractors Ltd
Theo McAllister
Graham Miller
Stephen Moate
Andrew Morland
Motokov UK Ltd
Jim Newbold

Renault Agriculture Ltd
Ivan Robinson
Malcolm Robinson
The Rural History Centre, Reading
Mervyn Spokes
Mike Stills
Robert H Tallman
The Tank Museum, Bovington
Volvo UK
Andy Watson

PICTURE CREDITS